INSTANT EGGHEAD

GUIDE: PHYSICS

D1530845

BRIAN CLEGG AND

SCIENTIFIC AMERICAN

INSTANT

EGGHEAD

GUIDE:

PHYSICS

ST. MARTIN'S GRIFFIN ⚬ NEW YORK

www.stmartins.com

Library of Congress Cataloging-in-Publication
Data

Clegg, Brian.
    Instant egghead guide. Physics / Brian Clegg
and Scientific American.—1st ed.
        p. cm.
    ISBN 978-0-312-59210-3
    1. Quantum theory—Popular works.
I. Scientific American, inc.   II. Title.   III. Title:
Physics.
    QC174.12.C546 2009
    530—dc22

                                        2009017043

First Edition: November 2009

10 9 8 7 6 5 4 3 2 1

CONTENTS

CHAPTER FOUR
RELATIVITY

INSTANT EGGHEAD

GUIDE: PHYSICS

CHAPTER

ONE

MATTER

STUFF

Physics, the topic of this book, is *the* fundamental science. The great physicist Ernest Rutherford once said, "All science is either physics or stamp collecting." He meant that most other science at the time was about collecting and categorizing information. Physics explained how the universe works.

In this section, the focus is stuff. What's this book—or you—made of? Is everything made of the same kind of stuff? Why is some stuff hard and some fluid? How do you turn one kind of stuff into another?

To the modern mind, stuff is obviously made up of atoms, assembled out of tiny components like a vast LEGO structure. Yet this isn't obvious. Look at a glass of water. Both the glass and the water inside seem to be continuous substance, quite unlike anything we're familiar with that's made up of small components. To understand matter we have to go beyond the obvious, to see in our minds what we can't experience with our senses, and that's part of the fun of physics.

ON THE FRONTIER

An atom is the single smallest particle of an element. It is as small as you can get and still have a bit of that substance.

Molecules contain more than one atom, joined together. This can be a pure element. For example, a molecule of oxygen contains two oxygen atoms, joined together. But it can also contain several elements, whether it's a simple molecule such as sodium chloride or the immensely long DNA chains that make up the complex molecules of life.

COCKTAIL PARTY TIDBITS

- ❖ One ancient Greek theory on matter suggested that you could cut stuff up into smaller and smaller pieces until eventually you could cut no more. What was left was uncuttable—in Greek, *a-tomos*—atoms.

- ❖ If you could squeeze all the matter in your body together, removing the empty space, it would pack into a cube less than a thousandth of an inch per side.

BROWNIAN MOTION

THE BASICS

Atoms are like small children: they are never entirely still. It's a remarkable contrast between the visible world and the world of the very small. Look at a glass of water. That water appears to be motionless, yet within the liquid the water molecules are frantically dancing.

In 1827 a Scottish botanist called Robert Brown was studying pollen grains of an evening primrose plant, suspended in a drop of water under a microscope. The tiny specks of pollen jumped about, constantly in motion.

There seemed to be no order to the motion, no rules for the way they moved. Instead the pollen grains' dancing was chaotic. This jerky dance was named Brownian motion, but remained little more than an oddity until Albert Einstein linked it to the behavior of atoms.

ON THE FRONTIER

Einstein produced three great papers in 1905. His works on relativity and the photoelectric effect get the glory, but his third paper on Brownian motion was just as significant. Until that time, the concept of atoms was entirely theoretical. But Einstein showed that the dance of the pollen grains was caused by the random impact of billions of water molecules.

Einstein used Brownian motion to show that the liquid the grains floated in was composed of many billions of gyrating molecules.

COCKTAIL PARTY TIDBITS

❖ When Robert Brown first saw Brownian motion he suspected it was the life source of a plant in action. It was only when he tried stone dust and soot and found the same effect with particles that were never alive that he confirmed that the size of the pollen grains was responsible for their motion.

❖ It wasn't until 1912 that French physicist Jean Perrin firmly established the existence of atoms. Until then, many scientists denied they existed.

ATOMIC STRUCTURE

It was surprisingly soon after finding proof that atoms existed that the name "atom" proved to be inaccurate. Even as Brownian motion was showing atoms and molecules to be real, it was becoming obvious that atoms weren't uncuttable.

A British scientist, J. J. Thomson, discovered in 1895 that there was a negatively charged particle within the atom, which would be named the electron. He imagined atoms were like a pudding with raisins spread through it. The raisins were the electrons and the rest of the pudding was positively charged, balancing this out so the atom had no charge itself.

ON THE FRONTIER

Thomson's pudding model was blasted apart by New Zealand scientist Ernest Rutherford. Rutherford had the idea of firing other particles into an atom and seeing how they reacted, like throwing a ball repeatedly at an invisible object and working out its structure by seeing how the ball bounces off. The ball he used was an alpha particle, the nucleus of the helium atom. These particles could be detected when they hit screens painted with fluorescent material. If the atom had been as Thomson imagined, the powerful alpha particles should plow through. Most did, but occasionally one bounced back. This unexpected result made Rutherford realize that atoms have a small, dense, positively charged core to deflect the alpha particles. He established the familiar idea of the atom being like a solar system with a positive nucleus at the center and negative electrons floating around it.

COCKTAIL PARTY TIDBITS

❖ The nucleus is so much smaller than the atom—10,000 times smaller than the whole—that it has been described as being like a fly in a cathedral.

❖ Remarkably, some scientists have seen an individual atom directly. In 1980 Hans Dehmelt of the University of Washington isolated a barium ion (an ion is an atom with electrons missing, or extra electrons added, giving it an electrical charge). When illuminated by the right color of laser light, the ion was visible to the naked eye as a pinprick of brilliance floating in space.

ISOTOPES AND NEUTRONS

The picture of the atom as a miniature solar system was initially simple. At its heart was a nucleus of positively charged protons providing 99.9% of its weight. Far out from the nucleus flew an equal number of negative electrons. Between them the charge balanced out. It wasn't clear how those protons, which should repel each other, stuck together, but otherwise the model made sense.

In 1932, however, another type of particle was discovered in the nucleus, about the same weight as a proton but with no charge: the neutron. Neutrons proved invaluable in explaining why there are different versions of the same element called isotopes, chemically similar, but weighing different amounts.

The chemical properties of elements are caused by the charged particles. But you can have versions of the same element with differing numbers of neutrons in the nucleus. This explains how chlorine, with 17 protons, can have an atomic weight (roughly the number of protons+neutrons) of 35.45. It doesn't have $18\frac{1}{2}$ neutrons; it's a mix of versions of the atom, some with 18 neutrons and some with 20.

Some isotopes are radioactive, unstable, and spew chunks from their nucleus to become another element. The best known is uranium 235. The numeral "235" is the atomic weight; it has 92 protons and 143 neutrons. When uranium 235 decays, it gives off several neutrons, converting the atom from uranium to thorium 231 (itself unstable). In a nuclear reactor or a nuclear bomb, those neutrons then blast into other uranium nuclei, triggering further fission. Of itself, uranium 235 decays slowly; it takes around 700 million years for half a chunk to decay. It's only with such a chain reaction that nuclear energy can be useful.

COCKTAIL PARTY TIDBITS

❖ There are 92 naturally occurring elements from the lightest, hydrogen, to uranium, the heaviest.

❖ Since 1972, the remains of 15 natural nuclear reactors have been found. Around 1.7 billion years ago a stable nuclear reaction took place in underground deposits of uranium. Because the amount of uranium 235 in the ground drops as it decays, it is unlikely such natural reactors would be found now.

FORGET THE PLANETS

If someone draws an image to represent an atom, chances are it will look like a solar system with the electrons neatly orbiting the nucleus. Just take a look at the logo of the International Atomic Energy Agency. The sad thing is, this model is wrong.

When you accelerate an electron it gives off bursts of light. Acceleration is a change in velocity (speed plus direction). The electron's speed would remain the same as it flew around the nucleus, but its direction would be constantly changing, which means it would be accelerating. Problem! Any orbiting electron should give off light, losing energy. It would hurtle into the nucleus to be destroyed, like a moth spiraling into a candle flame. This clearly isn't happening, or all our atoms would implode.

ON THE FRONTIER

Danish scientist Niels Bohr soon recognized this problem. He fixed it by putting his electrons on imaginary tracks. Instead of flying around the nucleus in any orbit, Bohr's tracks limited where an electron could be. Once on a track, the normal rules didn't apply: the imaginary track stopped energy leaking out. Electrons could jump from one orbit to another—giving out or absorbing a photon of light—but could not live anywhere in between. It wasn't possible for an electron to drift down and crash into the nucleus; they could only make leaps between fixed orbits. These jumps between different tracks, gaining or losing a quantum of energy with each jump, were called quantum leaps. Bohr had taken the atom digital.

COCKTAIL PARTY TIDBITS

❖ Such was the power of the image of the atom as a miniature solar system that some fiction writers envisaged it as literally true, with each electron a planet that could be inhabited.

ANTIMATTER

Fans of *Star Trek* know that the *Enterprise* is powered by antimatter, but antimatter is not fictional. It is like the familiar stuff that makes up our world, but every charged particle has the opposite charge. Instead of negative electrons, antimatter has positive positrons. Instead of positive protons in the nucleus, an anti-atom has negatively charged antiprotons.

Because it only differs in the placement of electrical charge, it's possible to do anything with antimatter that can be done with ordinary matter. You can build an anti-table or an anti-world. Antimatter has mass and behaves much as ordinary matter does. But don't expect to go out and buy some. Handling antimatter is tricky. When matter and antimatter get together, both are destroyed, converted into pure energy.

The simplest matter/antimatter reaction is when an electron and a positron collide. Their mass is converted into energy as two photons of light (gamma rays) according to Einstein's famous equation $E=mc^2$: the energy produced is equal to the mass of the particles multiplied by the square of the speed of light. Because of this tendency to annihilate, very little free antimatter is found.

There is still a big debate about why there is so little anti-matter. The Big Bang should have produced equal amounts of matter and antimatter, which then would wipe each other out, leaving a universe full of energy. That this didn't happen is usually explained by assuming that subtle differences in the properties of matter and antimatter meant that there was a little extra matter. As little as one particle in a billion may have survived the matter/antimatter wipeout. But that was enough.

Some have speculated that, instead, the universe became segmented, leaving vast pockets of antimatter, perhaps on the same scale as the observable universe.

COCKTAIL PARTY TIDBITS

❖ The amount of energy generated from matter/antimatter interactions is vast. One kilogram (2.2 pounds) of matter/antimatter would produce around 10^{17} joules (1 with 17 zeros after it). That's the energy output of a typical power station for six years.

QUARKS

Protons and neutrons are no longer considered fundamental particles. Each is made up of three smaller particles, quarks. There's a whole mess of quarks distinguished by characteristics known as flavors (no, really): charm, strangeness, top/bottom, and up/down. The proton is two ups and one down; the neutron two downs and one up.

Up quarks have a $2/3$ charge and down quarks $-1/3$, resulting in a positive charge of 1 for the proton and no charge at all for the neutron. We aren't used to nature coming up with quantities in thirds. But the unit of charge is arbitrary. We ought to say that up and down quarks have charges of 2 and -1 respectively—so a proton has a charge of 3 units—but because protons and electrons were the simplest particles known when the units were established we are stuck with thirds.

ON THE FRONTIER

No one has ever seen a quark, nor broken a proton or neutron into its components. It is difficult to do so, because the force that holds quarks together gets stronger as they move apart. Quarks were first predicted by a purely mathematical formulation of quantum theory. The existence of quarks themselves has since been indicated by experiments that show three constituents in a proton, and by the short-lived production of particles made up of combinations of different quarks.

COCKTAIL PARTY TIDBITS

❖ Although "quark" is usually pronounced to rhyme with bark, when American physicist Murray Gell-Mann came up with the name he wanted it to rhyme with dork. He says he used the "kwork" sound first without thinking about how to spell it, before coming across the line "three quarks for Muster Mark!" in James Joyce's *Ulysses*. This sounded apt, but Gell-Mann wanted to keep his original pronunciation.

SOLIDS

Of the three familiar states of matter, solids have the lowest energy; the atoms or molecules jiggle around less than in a liquid or a gas. In solid form, atoms or molecules link together. Although the solid is primarily empty space, the links between the particles give it a rigidity that distinguishes it from a liquid.

Solids come in several forms. Many are crystals. Here the links between the atoms or molecules form regular patterns, giving a tendency to shear on certain planes. Other solids are disordered. Disordered solids are often more flexible. Glass, for instance, although it can shatter, is a disordered solid, making glass fiber very flexible. There are also solids, often organic, with long chains of molecules linked together, making them particularly strong in one direction.

A single substance can have several solid forms. Carbon, for example, forms crystal planes that slide over each other in soft graphite, extremely hard interlocked crystal structures in diamond, or self-contained molecules shaped like a soccer ball, called buckyballs.

We are used to three states of matter—solid, liquid, and gas—but modern physics recognizes five. The most energetic is plasma, of which more later. The least energetic is a Bose–Einstein condensate, dreamed up by Albert Einstein in the 1920s. The Indian physicist Satyendra Bose had devised a way to describe light as if it were a gas. Einstein helped Bose firm up the math, but was also inspired to imagine a state of matter where intense cold or huge pressure forced it to share some of the characteristics of light itself. Such matter is a Bose–Einstein condensate.

COCKTAIL PARTY TIDBITS

❖ It's sometimes thought that glass is a liquid, because medieval window glass seems to have run down the panes, making them thicker at the bottom, but this merely reflects the way glass was made. Panes were uneven, and it made sense to put the thicker edge at the bottom.

LIQUIDS

In a solid, the bonds that link atoms or molecules are relatively rigid. But with increased energy (higher temperature) those bonds break. The particles are still attracted to each other, but without a rigid structure. The result is a liquid that can flow and adapt to the shape of a container.

Unlike gases, liquids form surfaces. On these surfaces, almost all the attractive forces between the liquid molecules are inward; this produces a skinlike effect called surface tension. Surface tension is why drops of water form, and why some insects can walk on water. When a liquid rests on a solid, the liquid is attracted to the solid by the same forces, resulting in wetting.

The liquid we are most familiar with, water, is atypical. The positively charged hydrogen in a water molecule is attracted to the negative oxygen in adjacent molecules. (This kind of attraction is called a hydrogen bond.) This bond gives water strange properties, such as expanding as it freezes.

ON THE FRONTIER

Fans of Kurt Vonnegut may have come across the concept of Ice Nine in his novel *Cat's Cradle*. Vonnegut described a form of ice that melts only at 114 degrees Fahrenheit (45

Celsius). If water ever got into an Ice Nine form, under normal weather conditions it would never get out of that form. Should a seed crystal of Ice Nine be dropped into a lake or an ocean it would spread uncontrollably, locking up the water supply and devastating the Earth.

Luckily, Ice Nine doesn't exist, although there is a type of ice with the intentionally similar name of Ice IX. This, however, isn't stable at room temperatures, and presents no danger to our water supply.

COCKTAIL PARTY TIDBITS

❖ Where a liquid meets the edge of a container it curves up or down, forming a meniscus. If the atoms or molecules in the liquid are more attracted to each other than the container they will curve down (as with mercury in glass); if they are more attracted to the container they will curve up (as with water in glass).

GASES

A gas, like a liquid, is a fluid, but because the atoms or molecules bounce around with significantly higher energy, their attraction to each other is overcome, so they don't form surfaces. A gas expands to fill the space available.

When a gas is inside a container, the molecules of the gas constantly collide with the side of the container. The combined result of this battering is that the wall of the container feels a force, the pressure of the gas. If you reduce the size of the container, gas molecules have less far to travel, so they hit the sides of the container more often. It turns out that pressure times volume stays constant. Discovered by the British chemist Robert Boyle in the 1660s, this is called Boyle's law.

It is also possible to increase the pressure by increasing the temperature, making the gas molecules speed up and batter the walls more. So increasing temperature in a fixed volume increases pressure; this is called Gay-Lussac's law after the French chemist Joseph Gay-Lussac who formulated it in 1809. This also means that, if you keep the pressure constant, then the volume of the gas has to increase as the temperature goes up, or decrease as the temperature goes down, a link known as Charles's law after the French scientist Jacques Charles who observed it in the 1780s.

These three laws combine to form the gas law describing the physical behavior of gases, saying the pressure times the volume, divided by the temperature, remains constant.

ON THE FRONTIER

Statistics is essential for dealing with gases. Statistics gives us an overview of a large body of items that we couldn't possibly hope to monitor individually. Almost any measurement on a gas in the real world (pressure, for instance) is statistical, combining the effects of the many billions of gas molecules.

COCKTAIL PARTY TIDBITS

❖ In air at room temperature the gas molecules move at around 500 meters per second, over 1,100 miles per hour. Luckily they are so light that even at this speed, the energy of each molecule is around 6×10^{-21} joules. That's 1/10,000,000,000,000,000,000,000th of the energy required to run a 60-watt light bulb for a second.

PLASMA

The fourth state of matter is one that everyone has experienced, but many don't know exists. It's plasma. This has nothing to do with blood plasma, the clear fluid part of blood. My dictionary defines a plasma as a gas containing ions rather than atoms or molecules. Let's not worry for a moment about those ions, but that definition is plain wrong. To call a plasma a gas is like calling a liquid "a gas with fluid properties." A plasma is more like a gas than a liquid, but it is still a different state of matter.

We've all experienced plasmas. The sun is a huge ball of plasma. Every flame on Earth contains some plasma too, although flames are relatively cool and thus are a mix of plasma and gas. A gas is what happens to a liquid if you heat it past a certain point; similarly a plasma is the outcome of heating a gas. As the gas gets hotter and hotter, the electrons around its atoms are bumped up to higher energy states. Eventually some have enough energy to fly off. These atoms lose electrons, and end up as a positively charged ion. Others easily gain electrons, sucking up spare electrons to become negative ions. This is a plasma.

Plasmas are common in the universe. Up to 99 percent of the universe's detectable matter is plasma. Although plasmas are gaslike in being diffuse and lacking surfaces, they are very different from gases. For instance, gases are usually good electrical insulators; plasmas are superb conductors.

COCKTAIL PARTY TIDBITS

❖ Unlike gases, plasmas can form structures because their charged particles respond to electrical and magnetic fields. The most common natural example of plasma structures on the Earth occur in the aurora and lightning.

THE PARTICLE ZOO

From the 1960s onward a whole range of new subatomic particles was discovered, a haphazard zoo of apparently unrelated particles. This has been rationalized, leaving us with the "Standard Model." It describes a collection of 24 different particles plus a 25th particle that's a loner.

The 24 divide neatly into 12 bosons and 12 fermions. Fermions are the particles that make up matter, and bosons transmit forces, enabling fermions to interact with each other. The boson we're most familiar with is the photon, and it's joined by W and Z bosons, various gluons, and the Higgs boson, thought to be responsible for giving other particles mass. (The 25th particle is also a boson, the graviton, a hypothetical particle responsible for gravity.)

In the fermion camp we find neutrinos, positrons, muons, and tau particles: particles produced when atoms are smashed together. More familiar are the electron and a whole mess of quarks, including the up and down quark to make those familiar neutrons and protons.

ON THE FRONTIER

The Higgs boson is yet to be spotted. The best hope to isolate it is at CERN (Conseil Européen pour la Recherche

Nucléaire), a vast international research establishment, straggling over (or, rather, under) the border between Switzerland and France near Geneva. Best known now for inventing the World Wide Web and its role in Dan Brown's thriller *Angels and Demons*, CERN is a place where the basic components of the universe are battered together with immense energy to analyze their makeup.

In the 84-kilometer-long Large Hadron Collider at CERN, particles are accelerated in a nightmare carousel, spurred on by vast magnets that take the power of a good-size city. When they are close to the speed of light, the particles smash into each other, causing an incredibly high energy reaction. The LHC is the first device capable of producing enough energy to make the Higgs boson visible.

COCKTAIL PARTY TIDBITS

❖ With almost zero mass and no electrical charge, neutrinos don't readily interact with other particles, and are difficult to detect as they fly past at near the speed of light. Around 50 trillion neutrinos from the Sun shoot through your body every second.

STRING THEORY

The Standard Model works well, but is complex, and can't deal with gravity. Enter string theory. According to string theory, every particle is made from the same fundamental building block, a string. These incredibly tiny components vibrate in different fashions, each pattern producing a different particle.

The nice thing about this model is that it is simple to grasp. But string theory comes at a price. The picture is simple, but the math is fiercely complex, and the theory needs the universe to have nine spatial dimensions.

It ought to be stressed that the strings in string theory aren't literally strings. The string description is a model. We can think of particles being strings, but really we mean abstract constructs that have behavior reminiscent of vibrating strings.

ON THE FRONTIER

M theory adds an extra dimension to string theory, making ten spatial dimensions. M theory has as a basic unit a "brane," a multidimensional membrane. This can have as many as ten dimensions, but in a one-dimensional form twisted through various other dimensions simplifies to a string (making string theory a subset of M theory).

M theory describes our universe as a three-dimensional brane, floating through higher dimensions of space. Its development proved a relief to many string theorists, as it unifies five different, incompatible versions of string theory.

COCKTAIL PARTY TIDBITS

- ❖ No one is sure what the "M" in M theory stands for. Membrane has been suggested (which makes sense) or mystery or magic; the physicist Ed Witten, who named M theory, has never come clean.

- ❖ Why don't we see all those dimensions? Some of them are thought to be curled up so small—smaller than an atom—that we can't detect them.

- ❖ These theories give no reason to choose that particular solution and make no new testable predictions. Because of this, some physicists argue that they aren't true science.

EXPANDING UNIVERSE

THE BASICS

When it comes to matter, you can't get any bigger than the universe.

The definitive discovery in understanding the universe was the realization that it is expanding. Back in the 1920s, the Russian Aleksandr Friedmann showed that general relativity, Einstein's theory describing how space and time interact, allowed for an expanding universe. Belgian scientist Georges Lemaître built on this to imagine tracking back an expanding universe to its beginning as a "cosmic egg" from which everything expanded. This theory was given support by American astronomer Edwin Hubble a few years later. He observed that distant galaxies were moving away from us; the whole universe seemed to be expanding. Given a few assumptions about the nature of space and time, and recent satellite observations, scientists have worked out from this expansion that the universe appears to have begun around 13.7 billion years ago.

ON THE FRONTIER

Einstein didn't like the idea of an expanding universe. He was more comfortable with the idea that the universe was static, a picture that was generally accepted before Hubble's work. So he fudged the relativity equations, adding in an arbitrary value called the cosmological constant, to counter any expansion and keep the universe fixed. He later called this constant his "greatest mistake," although modern theories require its use.

COCKTAIL PARTY TIDBITS

❖ We reach the Big Bang through time by tracking back the expanding universe, so it might seem possible to similarly track back spatially and say that a particular point is where the Big Bang happened. However, because it's space itself that's expanding, *everywhere* was originally that point where the Big Bang happened. There is no specific point where it all started.

DARK MATTER

The good news is that all the stars and planets we observe seem to be made of matter that works just like the stuff to which we are accustomed. But astronomers believe that things aren't that simple.

There's something wrong with the way galaxies move. If you spin something around very quickly, it flies apart. Most galaxies do spin quickly, and the only thing that stops the stars inside them shooting off into intergalactic space is gravity. But the best estimates of how much matter there is inside galaxies show it's too low to hold them together. Astronomers assume there is more matter out there that we can't detect, called dark matter.

There's plenty of stuff in the solar system that is difficult to see—asteroids, dust, and so forth—but even allowing for this magnified up to a galactic scale there is nowhere near enough matter. Even the massive black holes thought to be at the center of many galaxies, including our own, don't provide enough extra mass for this unexpected behavior. Dark matter has to be something more.

The existence of dark matter is given a theoretical boost by the results from the COBE and WMAP satellites, which measure the background radiation that fills the universe. Slight variations in this cosmic background are thought to show the accumulations of matter in the early years of the universe that later became galaxies. But the amount of matter that can be deduced from these maps suggests there wasn't enough ordinary matter to provide the gravitational impetus to form the galaxies we now see.

COCKTAIL PARTY TIDBITS

❖ Cosmologists divide possible sources of dark matter into MACHOS (MAssive Compact Halo Objects) and WIMPS (Weakly Interacting Massive Particles). MACHOS are conventional matter we can't see and WIMPS are some other kind of particles we haven't detected yet. Such hypothetical special particles would not interact with photons (so they can't be seen) but would still have mass, to help hold a galaxy together.

❖ Some cosmologists believe that gravity acts differently on the scale of a galaxy. This isn't as unlikely as it seems: after all, physics works very differently on the scale of atomic particles. A scheme called MOND (MOdified Newtonian Dynamics) explains the effects ascribed to dark matter without needing all that extra stuff.

DARK ENERGY

Dark matter isn't the only dark physics. Cosmologists also believe that there is something called dark energy. (Arguably this should be in the "Energy" chapter, but it is too closely linked to other aspects of cosmology.)

The Big Bang model of the universe describes an initial surge of inflation, followed by slower expansion. The battle between that expansion of the universe and the gravitational attraction of all the stuff in it should result in the rate of growth slowing. But by the 1990s we were better able to measure how the universe was expanding, and it is speeding up. Something is working against gravity to force the universe to expand faster. This "something" is called dark energy. Conveniently, this corresponds to Einstein's cosmic constant, although operating in the opposite direction.

ON THE FRONTIER

To compare rates of expansion now and in the past, the researchers who discovered dark energy had to peer far into the depths of the universe. Because light travels at a finite speed, the farther you look in space, the farther back in time you see. To judge distances, astronomers use "standard candles," which are objects in space they believe have a consistent brightness. The standard candle used here was a supernova, the vast explosion caused when a white dwarf star consumes a companion star and becomes unsustainably large. Because they are dependent on the size and type of star, such explosions are consistent in brightness, and so are ideal for looking way back in cosmic history.

COCKTAIL PARTY TIDBITS

❖ Dark energy is no delicate phenomenon. If it exists, dark energy is thought to provide 70 percent of the energy (and hence matter) in the universe. Remember that $E = mc^2$ means we can consider energy and matter interchangeable in terms of summing up the contents of the universe. More than two thirds of everything in the universe seems to be this strange source of energy that tears it apart.

THE BIG BANG

The lightest elements such as hydrogen have been around for the lifetime of the expanding universe, whereas the heavier elements making up solid matter were produced from lighter elements in early stars and supernova stellar explosions. But where did the matter come from in the first place?

The most commonly accepted view is the Big Bang, around 13.7 billion years ago. In this model, the Big Bang is the point where time and space came into existence. The modern Big Bang theory starts with a fluctuation in the newly created spacetime, an infinitely dense and hot speck that rapidly and vastly expanded in a sudden burst of inflation, before settling back to the more steady expansion that we now experience. What was initially an intensely hot fireball cooled and condensed, dropping from plasma to gas molecules, which clumped together under gravity to form the galaxies.

ON THE FRONTIER

It was once thought that the universe could run through a cycle of expansion and collapse, making the Big Bang just one step along a series of bang–expansion–collapse–crunch–bang cycles. However, there is no evidence that the universe will stop expanding, and there are difficulties with the physics of a simple repeated cycle, so the best bet under the basic Big Bang model is a single universe that will expand forever, becoming less and less energetic in a "heat death."

COCKTAIL PARTY TIDBITS

❖ Because all solid matter was produced in early stars and stellar explosions, Joni Mitchell got it right in her 1960s song "Woodstock": We are stardust.

❖ The term "Big Bang" was first used by astronomer Fred Hoyle in a BBC radio broadcast in 1950. Many think he was being sarcastic (he supported an alternative theory) but it stuck.

❖ The Big Bang is our best accepted guess of what happened, but it is based on indirect evidence. It is in no sense a well-proved theory, and is usually given a lot more weight than it deserves.

MULTIVERSES AND BOUNCING BRANES

THE BASICS

The Big Bang model is not alone as the suggestion of from where the matter in the universe came. The best evidence we have for the Big Bang comes from cosmic background radiation. This is a faint ripple of microwave electromagnetic radiation that fills all of space. Subtle variations in this radiation have been measured by the COBE and WMAP satellites, and data from these probes offer support to the Big Bang theory, but this isn't the only theory that works with the observed data.

One alternative keeps the Big Bang, but considers it a local event, rather than the beginning of all time and space. If this were the case it could be one of many Big Bangs in a multiverse of many universes, each being like a bubble blowing up within the much vaster spacetime of the cosmos. We would never be aware of the other bubble universes.

ON THE FRONTIER

Other alternative models for the universe are more exotic still. One, called the ekpyrotic universe, depends on M theory and imagines that our universe is one of a pair of universes, each on separate branes. Gravity is thought to be able to leak across from universe to universe, causing them eventually to collide, this collision providing the Big Bang. Because of their strange structure these bouncing branes could continue expanding indefinitely, occasionally colliding, providing an eternal cycle of existence with repeated Big Bangs and no specific point of creation, a universe that goes on forever.

There are other theories which suggest that our universe could be a virtual one, operating on a computer, like a grand-scale version of *The Matrix*. Alternatively, our universe could be a holographic projection that only appears to have three spatial dimensions, or could even be situated within a black hole.

COCKTAIL PARTY TIDBITS

❖ The first traces of cosmic microwave background radiation were picked up by an early radio telescope in Holmdel, New Jersey, in 1965. Researchers Robert Wilson and Arno Penzias first thought the hiss was caused by pigeon droppings on their instrument.

QUANTUM THEORY

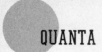

QUANTA

Quantum theory explains how the world works on the very small scale. We are used to seeing things in big lumps; our natural understanding of how things are relates to these "macro"-sized objects. But the rules of the macro world don't apply when you get down to the scale of atoms, electrons, and photons of light.

The word "quantum" (plural, "quanta") comes from the Latin *quantus* meaning "how much." In physics it is the minimum amount of a physical quantity that can exist. By extension, a quantum leap is the smallest possible jump something can make, usually the jump an electron makes from one level of energy to the next. This makes the way people speak about a major breakthrough as "a quantum leap forward" nonsensical.

The person who set the quantum revolution in progress was the German physicist, Max Planck. He realized that he could explain the way light was given off by hot things if he pretended that the light was tied up in little packets, which Einstein later called quanta. Planck didn't believe these packets existed; they just made his math work. It was Einstein who took the bold step of suggesting they were real things.

All electronics depends on quantum theory, but in the future we could see more dramatic quantum devices. In computing, for example, there are some problems that would take the lifetime of the universe for the most powerful of today's computers to solve. If we can use individual quantum particles as bits in a computer, that computer could use a quantum particle's capability of being in more than one state at once to smash through these problems.

For example, if you try to look up someone's name in a telephone book, given a number, it can take a long time. With a million entries, you might have to check a million numbers before you got to the right one. A quantum computer would pin down the right entry with just 1,000 checks.

COCKTAIL PARTY TIDBITS

❖ When Max Planck was in college, he had difficulty choosing between physics and music. His physics professor, Phillip von Jolly, recommended going for music, as physics was pretty well fully explained, apart from a couple of small problems. Those small problems resulted in quantum theory and relativity, the two central theories of all modern physics.

THE ULTRAVIOLET CATASTROPHE

Nineteenth-century physics (the sort most of us learned in high school) had a few issues that remained to be sorted out. The one that ushered in the quantum age was "the ultraviolet catastrophe."

This was a problem with blackbody radiation. A blackbody is a hypothetical object, so black it absorbs every bit of light that's sent into it. Every single photon hitting a blackbody disappears. It's truly black, a visual void. Blackbody radiation is light emitted by a blackbody. Yet this seems not to make any sense. We've already said that it absorbs every bit of light that hits it, so how can it radiate?

This is because there are two separate reasons that matter gives off light. One is reflection. But the other is generating photons of its own. As you heat up an atom you push energy into its electrons, and every now and then, one of those energized electrons pumps out a packet of energy in the form of a photon. The object glows. The hotter the object, the more energy the photons have. As something gets hotter it glows red, then yellow, then white. Blackbody radiation is this kind of glow. And because a blackbody doesn't let out any light that hits it, there's no confusion from reflected light.

It was known that in radiation from a blackbody, the

higher the frequency, the more energy should be emitted. This meant for high-frequency light such as ultraviolet, a blackbody should absolutely pour out radiation, heading off to an infinite amount at the very high frequencies. This clearly didn't happen.

German physicist Max Planck got around this problem by imagining that the light from a blackbody was bundled up into packets. In 1900 he showed that for packets of light there are only a limited number of ways the high-frequency light can be emitted.

❖ Planck was awarded the Nobel Prize for this idea, even though he only half believed in it. His citation says that it was "in recognition of the services he rendered in the advancement of Physics by his discovery of energy quanta." Planck was lucky to receive the prize during the First World War; that year there were no prizes awarded in literature, medicine, or peace.

EINSTEIN AND THE PHOTOELECTRIC EFFECT

THE BASICS

In a remarkable paper written in 1905 (the paper for which he later won his Nobel Prize), Albert Einstein suggested that light was made up of quanta. Instead of being continuous waves, he imagined it to be divided into the minute packets of energy envisaged by Max Planck, really divided up into particles (later called photons), instead of being waves as everyone thought.

In his paper on light, Einstein also showed that the radiation inside a blackbody behaves just as does a gas of particles; he could apply the same statistical techniques that he had already successfully applied to gases. What's more, Einstein predicted this quantized light should generate a small electrical current when shone on certain metals, something that had been observed but had yet to be explained. This photoelectric effect clinched the paper's significance.

ON THE FRONTIER

Einstein's idea was absorbed and amplified by the man who later became Einstein's chief sparring partner over quantum theory, the Danish physicist Niels Bohr.

In 1913, Bohr devised a model of the atom's structure that relied on Einstein's quanta to explain its workings. His idea was to consider the atom, with its tiny but heavy central nucleus surrounded by much smaller electrons, as if it were a sun with its attendant planets in orbit. As mentioned in Chapter One, this would only work if the electrons were forced to travel certain fixed paths. Instead of existing anywhere around the nucleus, they were forced to jump from orbit to orbit, and each jump would result in absorbing or giving out a packet, a particle of light: a photon. Bohr made quanta central to the interaction of light and matter.

COCKTAIL PARTY TIDBITS

❖ Planck criticized Einstein's ideas in a condescending fashion. When Planck recommended the younger man for the Prussian Academy of Sciences in 1913, he asked that they not hold it against Einstein that he sometimes "missed the target in his speculations, as for example, in his theory of light quanta. . . ."

WAVE/PARTICLE DUALITY

Einstein's bold move of describing light as particles caught everyone by surprise, because if there was one thing everyone was certain about, it was that light was a wave.

Thomas Young showed in a beautifully simple experiment in 1801 that light could produce interference patterns when it passed through a pair of narrow slits. The mingled beams threw fringes of light and dark onto a screen, corresponding to the addition and subtraction of the ripples in light waves, just as waves interacted on the surface of water. No other explanation seemed possible.

Scientists of the time could not comprehend how these patterns could be developed by a stream of particles. A particle had to follow a single path from source to screen. Passing particles through a pair of slits should result in two bright areas (one behind each slit) and large swaths of darkness, not the repeating dark and light patterns that everyone could clearly see. Similarly, light, like waves, can bend around corners, a phenomenon called diffraction, whereas particles are limited to bulletlike straight lines.

So how was science to cope with this new discovery that light behaved as if it were a particle when it interacted with matter? The answer was to say that light had both wavelike and particlelike properties, so-called wave/particle duality.

These ideas on light are models. Not literal models like the ball-and-stick molecule models you might have played with in high school, but mental models. When we say light is a wave or it's a particle, we mean that we're using the model of a wave or a particle to explain its behavior. Light is *like* a wave or particle, but these are both big, human-scale world things. In the quantum world, light is just light.

COCKTAIL PARTY TIDBITS

❖ In 1924, the magnificently named Duke Louis de Broglie thought that if light particles could behave as waves, why not other quantum particles as well? He showed that electrons, normally considered particles, could also behave as waves, producing interference patterns and being diffracted.

PROBLEMS WITH PROBABILITY

The German physicist Werner Heisenberg followed up de Broglie's discovery that electrons could act as waves by producing a totally abstract mathematical description of quantum processes. In parallel, the Austrian Erwin Schrödinger worked on the math of de Broglie's waves to describe how the waves of quantum particles changed over time. These two apparently conflicting views were pulled together by Paul Dirac to make quantum mechanics. But there was a problem.

If Schrödinger's wave equations were literal descriptions of the behavior of quantum particles then particles should self-destruct. If a particle such as an electron were literally a wave, following the behavior specified by Schrödinger's equations it would spread out in all directions and dissipate. The solution to making these wave equations usable came from Einstein's friend, Max Born.

To make Schrödinger's wave equations match the observed world Born suggested that they did not describe how a particle moves, but rather the *probability* that a particle would be in a particular place. The equations weren't a description of a particle, but a fuzzy map of its likely locations.

This basis of quantum theory on probability was a problem for Einstein. He could not accept this randomness, feeling there should be a strict causal process underlying physics. As far as he was concerned, if an electron jumped out of a piece of metal, the time it occurred and the direction it traveled in could have been predicted had all the facts been available. Quantum theory said it wasn't possible to know either time or direction in advance. Similarly, quantum theory assumed that a particle didn't have a position until a measurement was made: the act of measurement transformed its position from a probability to an actual value.

COCKTAIL PARTY TIDBITS

❖ The probability-based nature of quantum theory is apparent in the measure of radioactive "half-life." This is the time when, on average, half of the unstable atoms in a radioactive material will have decayed. But we can't say which atom, or when.

THE UNCERTAINTY PRINCIPLE

Heisenberg showed that quantum particles had pairs of properties that were impossible to measure simultaneously in absolute detail (properties are just aspects of an object such as
its mass, position, or velocity). The more accurately you know
one of these properties, the less accurately you can measure
its partner. For example, the more a particle's momentum is
known, the less you can say about its position. (Momentum is
the mass of the particle multiplied by its velocity.) At the ex-
treme, if you know exactly what momentum a quantum particle has, it could be anywhere in the universe.

The uncertainty principle makes it necessary for quantum particles to be always on the move. If they stopped you
would know exactly where they were and exactly how big
their momentum was (zero). The reason there is an un-
reachable minimum temperature, absolute zero (−273.15°C
or −459.67°F), is that temperature is a measure of the energy of movement of the atoms in the body being measured.
At absolute zero, they would stop.

ON THE FRONTIER

To get the idea of the uncertainty principle, imagine taking
a photograph of an object that is flying past at speed. If you

take the picture with a very quick shutter speed it freezes the object in space. You get a clear image of what the object looks like. But you can't tell anything about its movement from the picture. It could be stationary; it could be hurtling past. If you take a photograph with a slow shutter speed, the object will show up on the camera as an elongated blur. This won't tell you a lot about what the object looks like, but will give a clear indication of its movement. The tradeoff between momentum and position is a little like this.

COCKTAIL PARTY TIDBITS

❖ When Heisenberg came up with the uncertainty principle he used the example of an imaginary microscope looking at a quantum particle. As the photons of light used to look at the particle hit it, they would move it, so its position was impossible to determine exactly. Heisenberg was reduced to tears when his boss Niels Bohr pointed out this was rubbish; the uncertainty was inherent in the particle, regardless of whether you looked at it.

TUNNELING

Imagine a particle floating along until it hits a barrier, something it can't pass through. Because the location of a quantum particle is spread out across space with different probabilities, it's possible that a particle is already on the other side of the barrier. This is a phenomenon that is frequently observed. When a quantum particle passes through a barrier this way, it is said to have tunneled through. Unlike a conventional tunnel, the particle spends no time in the barrier, appearing instantly on the other side.

In principle any quantum particle can tunnel through any barrier. All the quantum particles in a car could simultaneously tunnel through the wall of the garage and appear on the other side. But the chances of this happening are so small, you could wait the entire lifetime of the universe and never see it happen. Taken at the level of individual quantum particles, however, this effect happens all the time.

A barrier can be anything that the particle hasn't got enough energy to get through. It could be something a particle would normally be absorbed in or reflected from, or it could be something that repels the particle.

Tunneling can be used to send a signal faster than light (see Chapter Three, the section "Faster than Light"), as the particle spends no time while tunneling. Imagine a photon of light crossed an inch of space, then an inch of barrier that it tunneled through, and then another inch of space. It covered three units of distance in the time light normally takes to cover two, so it traveled at one and a half times the speed of light.

COCKTAIL PARTY TIDBITS

❖ Without quantum tunneling we wouldn't be alive. The Sun works by forcing the nuclei of hydrogen atoms so close together that they fuse to make helium, releasing gouts of energy. But even the vast temperature and pressure of the Sun isn't enough to cram the positively charged nuclei close enough against the barrier of repulsion. It's only tunneling that lets them jump closer and fuse. It's very unlikely, but there are so many particles in the Sun that millions of tons of hydrogen convert to helium every second.

SUPERPOSITION AND QUANTUM CATS

There is an absolute difference between the "macro" world and the world of the quantum particle. It's painfully present in the phenomenon called superposition. Imagine a simple macro-world object with two possible states: a tossed coin. It could be heads or tails with a fifty-fifty chance. But until I uncover the coin we don't know which value it has. The fact remains, though, it *has* a value.

Contrast this with a property of a quantum particle that can have two different values, with a fifty-fifty chance. Until I make a measurement, the particle has both values at once. It is, in coin terms, both heads and tails. This is a superposition of states. When I make the observation, it randomly collapses to one value or the other. But until then the value isn't selected.

It might seem that there is no way of distinguishing between something having a value that's hidden and something that has both values until observed, but there are plenty of quantum experiments that provide different results for the two, and they all demonstrate that the quantum particle really is in both states at once.

This ability to be in more than one state is demonstrated by Young's slits. In the original experiment, light is shone through two narrow slits, forming dark and light fringes as parts of the light reinforce or cancel out. But the experiment works even if you send individual photons of light through, one at a time. The photons go through both slits at once and interfere with themselves. If you check which slit a photon goes through—making an observation and collapsing the superposed states—the interference fringes disappear.

❖ It was the idea of superposition that inspired Erwin Schrödinger to come up with the thought experiment called Schrödinger's cat. He imagined a cat in a box with a cyanide cylinder triggered by the value of a quantum particle's property. One value would leave the cat alive, the other would kill it. As the particle would be in a superposition of states before the value was observed, until the box was opened, wouldn't the cat be both dead and alive?

❖ It's generally thought that Schrödinger's cat isn't a problem because it requires a device in the box that makes a measurement on the particle, and that would collapse the particle into a single state. No fun for the cat, though.

THE COPENHAGEN INTERPRETATION

It sounds more like *The Bourne Identity*, but the Copenhagen interpretation is a part of quantum physics. Quantum physics works stunningly well. Quantum electrodynamics (QED), for example, the science of how matter and light interact, makes the most accurate predictions in all of science.

But quantum physics also has its dark side. Scientists don't just ask, "What?"; they also ask, "Why?" They want to know *why* quantum physics works the way it does, in a way that seems unreal in our familiar macro world. The most widely accepted answer is the Copenhagen interpretation.

The Copenhagen interpretation is not a document; it's more the collected views of Bohr's group of scientists. It includes the idea that the behavior of a quantum particle is subject to a probability equation called the wave equation, the uncertainty principle, wave/particle duality, and the idea that as systems of quantum particles get larger they become closer to classical (normal macro) systems. It also establishes that quantum particles are in a state of superposition until observed.

❖ The Danish capital, Copenhagen, features in the interpretation because Danish scientist Niels Bohr, and his institute in Copenhagen, were central to the debate over what lay behind the math.

❖ If interpretations of quantum theory have your head spinning, take comfort from the great American physicist, Richard Feynman. He said, "The theory of quantum electrodynamics describes Nature as absurd from the point of view of common sense. And it agrees fully with experiment. So I hope you can accept Nature as She is—absurd. . . . Please don't turn yourself off because you can't believe Nature is so strange."

MANY WORLDS

Although the Copenhagen interpretation is the most widely held understanding of quantum theory, there are a number of alternatives, of which the most important is the many worlds interpretation.

This comes in two different flavors, and is a simple theory with complex consequences. The idea is that each time a quantum particle has two possible options, the universe changes. In one flavor of "many worlds," each quantum decision flips the universe down a different path. It's as if the universe were a ball in a vast pinball machine, and each of the trillions of quantum decisions being made results in a different bounce of the ball off a bumper, sending it flying in a new direction. In the second flavor, there's a separate universe for every single state of every particle that ever has and ever will exist. When we observe a particle and see a particular value, we jump to a different, already existing universe, like a railroad car passing over a switch in the rails.

ON THE FRONTIER

British scientist David Deutsch has suggested that we could test whether the many worlds interpretation is true. To do this, we would have to build a computer that was self-aware enough to know how it "felt" about observing the state of a quantum particle. If the Copenhagen interpretation were true, such a computer would only know about the one value of the state that was observed. But in the many worlds universe, the computer would see the different possibilities interfering with each other. Even Deutsch, however, has no suggestion on how to build such a computer.

COCKTAIL PARTY TIDBITS

❖ One of the wackier interpretations held by serious physicists is John Wheeler's participatory anthropic principle. This variant of the Copenhagen interpretation reckons that the superposition of states of a quantum particle only collapses to a specific value when observed by a *conscious* individual. In such a universe parts of the universe that aren't consciously observed remain in a superposed state. As Einstein observed to ridicule this theory, if it is true, the Moon isn't there unless someone is looking at it.

ELECTRONICS

It can seem that quantum theory has little relation to the everyday world, yet quantum physics still has a huge influence, and nowhere more than in the field of electronics.

Electronics is inherently quantum. Even crude vacuum tubes involved manipulating electrons, quantum particles. The simplest vacuum tubes consist of a heater, such as a light bulb filament, and a positively charged wire, both in a vacuum inside a glass tube. The heater gives off electrons, which are attracted to the positive wire, so there can be a flow of electrical current in that direction. However, the reverse doesn't happen as the positive wire isn't giving off electrons. The result is a diode, a one-way flow valve for electricity.

More complex tubes have three metal components: the third is a grid between the other two. Depending on how this is charged, it controls the flow between the other two electrodes. This is an amplifier. A small change in current in the grid controls a big change in current flowing across the tube. This "triode" tube has the same function as the solid-state transistor that made modern electronics from the PC to the iPod possible.

The solid-state electronics that make the electronic devices that fill our homes possible go further, using many aspects of quantum strangeness, such as the ability to tunnel across barriers. When designing such electronic devices, quantum physics is brought directly into play.

COCKTAIL PARTY TIDBITS

❖ Physics crops up throughout the workings of an electronic device such as an iPod. The display, for instance, is liquid crystal. This uses polarization of light to display an image. Light from a backlight passes through two polarizing filters. These are rotated at 90 degrees to each other, so no light emerges. But the liquid crystal placed between the filters rotates the light's polarization by an amount that's controlled by the electrical current across the crystal. The result is the ability to control just how much light gets through, and, with millions of tiny pixels, build up a complex picture.

ENTANGLEMENT

THE BASICS

Einstein never liked quantum theory and spent a fair part of his life dreaming up hypothetical experiments to disprove it. One of these, published in 1935, showed that if quantum theory were true, it would be possible to separate two specially linked particles to opposite sides of the universe and a change in one would be instantly reflected in the other. This contradicted Einstein's idea that nothing could communicate faster than light, apparently exposing a flaw in quantum theory.

By the 1970s a French researcher, Alain Aspect, was able to show that Einstein was wrong and this "spooky connection at a distance" really did exist. Called quantum entanglement, it is most easily demonstrated using a property of quantum particles called spin (not actually about spinning around) which when measured will always have one of two values, up or down. Measure the spin of one entangled particle— say it turns out to be "up"—and instantly the other particle is "down." Yet until the moment of measurement, neither particle had a value for this property.

Entanglement can't be used to communicate faster than light. The properties measured can't be controlled, so you can't force the outcome and send a signal. But entanglement can be used for some dramatic applications, from super-fast computers to teleportation.

Austrian scientist Anton Zeilinger leads world experimentation on entanglement. He has already demonstrated using entangled particles securely across five miles in Vienna. This is an important distance because it is the equivalent of sending entangled particles to a communication satellite in geostationary orbit.

COCKTAIL PARTY TIDBITS

❖ Using entanglement to encrypt information means that the key is totally secure, generated directly by the random effects of entanglement, and should the message be intercepted the entanglement will break, alerting the parties to the security breach.

❖ Entanglement has been used to produce a miniature version of a *Star Trek* transporter, transferring the properties from atoms in one place to atoms in another, making the second set identical copies of the first while destroying the originals.

QED

The physics of the interaction of light and matter is called quantum electrodynamics, QED for short. The work of American physicists Richard Feynman and Julian Schwinger, independently duplicated by Japanese scientist Sin'Itiro Tomonaga, this is a remarkable theory that explains how much of the world works.

Feynman's secret weapon was his visual mind. Rather than just use equations, he liked diagrams. Feynman developed a set of diagrams covered in little arrows, where the size of the arrow indicated the chance of a particular event happening and the direction of the arrow indicated the point in time, so the arrows rotate with time as does the second hand of a clock. By combining all the arrows for the possible ways a photon could behave he could accurately predict its behavior.

The more he thought about these little arrows of probability, the more Feynman could see that all the behaviors of light—reflection, refraction, interference, diffraction—that seemed to require waves could be explained purely in terms of photons.

Richard Feynman said, "I want to emphasize that light comes in this form—particles. It is very important to know that light behaves like particles, especially for those of you who have gone to school, where you were probably told about light behaving like waves. I'm telling you the way it does behave—like particles." We use waves at school because they're easier to understand, not because they best describe the behavior of light.

COCKTAIL PARTY TIDBITS

❖ QED changes the way we look at some simple actions of light. When light reflects off a mirror, for example, it doesn't hit the surface and come bouncing back the way a ball does. Instead, photons of light are absorbed by the electrons in the atoms of the mirror's surface and new photons are emitted, causing the reflection.

QUANTUM REALITY

Some physicists believe that quantum effects apply to the very nature of space and time. A quantum universe is a digital universe. You can imagine space divided up into incredibly small pixels, like the dots on a computer screen, but much smaller. Digital measurements are made up of a series of discrete values, but analog is continuous.

This isn't just a vague hypothetical notion; we know just how big these quantum pixels are. Each side is a Planck length. This minimum distance is incredibly tiny, around 1.6×10^{-35} meters (that's zero followed by 33 zeroes then 16). By comparison, a proton is huge, more than 1,000,000,000,000,000 Planck lengths across. Because of the uncertainties in quantum theory it is impossible to measure anything smaller than a Planck length.

Some believe that this is the inherent granularity in the universe, in effect, that the universe is made up of Planck-length granules, and there is nothing beneath that length. Others think that it is a limit on measurement, but that reality is continuous.

Combining the Planck length with light speed produces a minimum unit of time, the time it would take light to move a Planck length. If the universe is physically broken up into Planck length "pixels" then there is no meaning for a unit of time smaller than the time it takes light to skip from one pixel to the next. The time unit is even smaller than the Planck length, around 5×10^{-44} seconds. Some argue that these are quanta of time, digital time.

COCKTAIL PARTY TIDBITS

❖ Physicist David Bohm came up with an alternative to the Copenhagen interpretation that did away with the need for the probability-based physics that Einstein hated, but required a very strange quantum reality. For Bohm's interpretation to work, the concept of locality had to be dropped. This meant that quantum particles are everywhere in the universe simultaneously. Distance would no longer be a fundamental property, just a concept that emerged from the way we interact with particles. Everything becomes part of a complex whole.

BLACK HOLES

Although quantum physics is mostly about the very small, there are some hypothetical large-scale quantum objects. One was the universe at the point of the Big Bang. Another is a black hole.

Black holes were dreamed up in the 1700s when British astronomer John Michell imagined escape velocity (the velocity needed to get away from a planet) getting bigger and bigger as the planet got more massive. With a heavy enough star, Michell realized, the escape velocity would be bigger than the speed of light. The result would be a dark star, or as the astronomer John Wheeler first called them in 1969, a black hole.

The modern idea of a black hole came from Einstein's general relativity, which considers gravity to be a warp in space. The more massive a body, the more it bends space. With enough mass in a small enough volume, space would bend so far that nothing—light included—would get out. To get the Sun, a middling-size star 1.4 million kilometers across, compressed enough to go black it would have to be condensed to just 3 kilometers in diameter.

Normally when a star is active, the outward pressure from the nuclear reactions that power it keeps the star "fluffed up," but as nuclear fuel runs low, pressure drops and the star collapses. Now another force comes into play, the Pauli exclusion principle which means similar particles of matter that are close in distance must be different in velocity. This counters gravitational collapse, unless the star is too massive. The mass required for this is around one and a half times that of the Sun. Some such stars explode as a supernova. Otherwise, the star should contract, getting smaller and smaller until it becomes a black hole. In theory, the contraction continues until there is a singularity, a quantum point of infinite density, at the center of the black hole.

COCKTAIL PARTY TIDBITS

- If you flew toward a black hole, the difference in gravitational pull between your feet and your head would stretch you out long and thin like a piece of spaghetti.

- General relativity says the stronger the gravitational pull, the slower time runs as seen from the outside. If we watched an object traveling into a black hole it would get slower and slower before stopping forever at the event horizon (the point beyond which no light escapes).

CHAPTER

THREE

LIGHT

THE MECHANICS OF SIGHT

Light's most obvious role is enabling us to see. In the seventeenth century, René Descartes demonstrated the optics of the eye by scraping the back off a bull's eye to reveal a hazy projection of the view through the lens. But the process of seeing is nothing like using a camera. Your brain doesn't take in a whole picture. It makes up an illusory image by combining the input of various modules that handle shapes, movement, edges, and more.

The retina at the back of the eye contains around 130 million light-sensitive receptors. When a photon of light reaches one, it triggers a photochemical reaction. This generates a signal that feeds information through the optic nerve to the brain. This nerve has considerably fewer fibers than there are receptors in the eye; the signal has already been processed before reaching the brain.

The combined image you "see" is an illusion, generated from the stimuli on the optic nerves. This becomes obvious when you consider how steady your view appears. In practice, your eyes are rarely still, undertaking tiny jerking movements called saccades. If you took a true "cameralike" view, everything would constantly twitch. Your brain shows you what you need to see, not a true image.

It's often said that we see moving images from the stream of still pictures emerging from a movie projector or a TV because of "persistence of vision." Modern research has shown that it couldn't work. The still pictures aren't projected long enough to form a persistent afterimage, and anyway such persistence would result in a blurred mess, not movement. We see moving pictures because of the way our brains construct an imaginary view of the world from the individual data elements provided.

COCKTAIL PARTY TIDBITS

- ❖ Our eyes are so sensitive that we can see a candle flame ten miles away on a truly dark night.
- ❖ Human sight is very flexible. Outside on a sunny day it's 100 times brighter than in a typical office, but our eyes balance out the difference. Full moonlight is around 300,000 times weaker than sunlight.

EARLY IDEAS OF LIGHT

The Greek philosopher Euclid, working around 300 BC, was the first to suggest that rays of light travel in straight lines. He still believed earlier ideas that we see because of light streaming from our eyes, but he had transformed light from a diffuse vaporous phenomenon to something that traveled in straight lines, its behavior predictable by the newfangled geometry.

This straight-line concept was used by Arab philosophers such as Alhazen, born in 965 AD. He discarded the idea of fire from the eyes, and developed the idea of rays of light from a source such as the Sun, reflected by mirrors, and bent by refraction as it passed from material to material.

ON THE FRONTIER

By medieval times, scholars such as the thirteenth-century friar Roger Bacon developed more detailed ideas on light. The Greek philosopher Aristotle had argued that light was infinitely fast, but Bacon argued it took time to travel, like sound but much quicker. He devised a mechanism called multiplication of species, where an object giving off light produced a series of pulselike phenomena, called species. Each of these would then produce more species, fanning out until the light reached the observer. He also drew detailed diagrams of light rays passing through lenses and mirrors and worked out how the rainbow could be formed by the refraction and reflection of sunlight within drops of rain.

COCKTAIL PARTY TIDBITS

❖ Alhazen is said to have pretended to be mad to save his life. He was employed by the Egyptian King al-Hakim to work out how to control the flow of the Nile. When Alhazen found this wasn't possible, he realized that his life was at risk. Rather than admit failure, he pretended madness, keeping up this pretense for years until the king died.

THE ELECTROMAGNETIC SPECTRUM

Although we often use the word "light" to mean visible rays, it is part of a much wider range of electromagnetic radiation. The first clues as to what this radiation might be came from the work of self-made English scientist Michael Faraday in the early 1830s, when he showed that a moving electric wire produced magnetism and a moving magnet produced electricity. He went on to speculate that light might be some form of electrical or magnetic vibration.

Spurred on by Faraday's observations, Scottish scientist James Clerk Maxwell realized that if you allowed magnetism and electricity to interact at one special speed, and that speed only, then the electricity produces magnetism, which produces electricity and so on, hauling itself up by its own bootstraps. That speed was the speed of light. It was a perfect, self-sustaining perpetual motion machine.

Maxwell defined light by the forces that produce it, rather than the eye's ability to detect it. This move away from a dependence on the eye was echoed by new "colors" that were being discovered, extending the spectrum beyond the visible limits of red at one end and violet at the other. The light that our eyes respond to occupies a tiny segment near the middle of this huge span of electromagnetic radiation. If the whole range of light, visible and invisible, were represented by a rainbow, we would see only a thin slice out of the green segment.

ON THE FRONTIER

The spectrum of light varies with energy, or if you consider light a wave, with the frequency of the wave. At the lowest energy end is radio, including TV, cell phones, and wireless Internet. Then come microwaves, infrared, visible light, ultraviolet, X-rays, and gamma rays. This most energetic form of light is produced by radioactive material and is the most dangerous in the degree of damage it can do to the human body.

COCKTAIL PARTY TIDBITS

- Infrared was discovered by German/English astronomer William Herschel in 1800. He was observing how a thermometer bulb was warmed in different parts of the spectrum from a prism. Not sure exactly where the red stopped, he moved the thermometer outside the visible area and was shocked to discover that light continued beyond the red, with even greater heating capacity.

- Faraday was not known for his sense of humor, but it's said that when British Prime Minister Robert Peel asked him what use his new invention of the electrical generator was, Faraday replied, "I know not, but I wager one day your government will tax it."

COLOR

When Isaac Newton took home a prism from the Stourbridge fair in 1664, this triangular block of glass was considered nothing more than a toy to produce miniature rainbows. It was assumed at the time that the rainbow colors were caused by impurities in the glass. But Newton sent a tiny sliver of the spectrum of light from one prism through a second. If the glass tinted the light, the second prism ought to have changed the shade, but it didn't. Newton had demonstrated that the colors of the rainbow were already present in white light.

There is often confusion over the primary colors from which all other colors can be made. The true primary colors, the primaries of light, are red, blue, and green. By mixing lights of these shades it's possible to generate all the other colors. The appearance of color in things, whether pigments or objects, is down to the negative versions of these colors. These form the primaries for pigment: cyan, magenta, and yellow, often confusingly described as blue, red, and yellow.

This was also something that Newton untangled. When white light falls on an object, some of the colors are absorbed, and others are reemitted. When we see something as red, all the colors of light except red have been absorbed by the object. Because objects appear to have a color by subtracting some of the light, the result is an inverted set of primaries.

Human eyes have four types of sensor: rods, sensitive to black and white, and three types of cones, which handle a range of hues, each type responding most to one of the primary colors. At night, the more sensitive rods are the only sensors active; night vision is monochrome. But there is a cutover period (called mesopic vision) when both types of vision occur together. It's as if there were a whole new color added to the spectrum that hadn't existed before. Sight at this dusky light level has strange qualities, perhaps why so many ghosts are seen at dusk.

COCKTAIL PARTY TIDBITS

❖ Some birds have an additional color receptor in their eye that enables them to see ultraviolet. This is how a hovering hawk detects a small rodent by the roadside. The brown animal is invisible, but its urine trail clearly stands out in ultraviolet.

LIGHT WAVES

Isaac Newton thought that light was a stream of particles he called corpuscles, but this view was overtaken by the wave hypothesis, considering light to be like the ripples on a still pond after a stone is dropped in it.

Light's identity as a wave seemed certain after experiments done by English scientist Thomas Young in the nineteenth century, which showed two light beams interacting just as waves do. Once light was seen as a wave, the colors of light had a sensible explanation. The key characteristics of a wave are its velocity, its wavelength, and its frequency (velocity equals wavelength times frequency, so given any two, you can work out the third). The wavelength is the distance the wave travels in getting from a particular point in its cycle (say its highest peak) to that same point in the next cycle. The frequency is the number of complete wavelengths it cycles through in a second.

All light seemed to have the same velocity in any particular substance, but different frequencies corresponded to different colors.

A real puzzle was what light was a wave in. When a wave travels through water, the molecules of water move: no water, no wave. If you put a bell inside a glass jar and suck out the air, you can't hear the bell because there is nothing for the sound waves to travel through. But you can still see the bell, and you can see stars across empty space. It was decided that everywhere was filled with a substance called the ether. This was so diffuse that you couldn't detect it, yet it had to be perfectly rigid, because light didn't lose energy as did a normal wave. The ether was finally disproved in 1887 by American physicist Albert Michelson.

COCKTAIL PARTY TIDBITS

❖ Initially, many thought that light was like sound. Sound is a compression wave (or longitudinal wave). This means that as it moves along it squashes up the material it's moving through, then expands, squashes and expands, like a concertina. However, it was later established, because of the effect called polarization, that light is more like a wave traveling down a string, a side-to-side or lateral wave, oscillating at right angles to the direction in which it's traveling.

PHOTONS

By the early years of the twentieth century, light was being described as packets of energy, particles that were given the name "photon" by American chemist Gilbert Lewis. This made it much easier to accept that light could pass across empty space with no medium to travel through.

It wasn't until quantum electrodynamics (QED) was developed that it was possible to explain a problem that Newton himself discovered. This was the problem of the beamsplitter, which allows some light to pass through, but sends the rest back the way it came. We all have beamsplitters at home: the windows. At night, light passes out through the window; you can see in. But look at the window from inside the room and it's almost like a mirror. Much of the light reflects back off the glass.

Newton wondered how a particular photon knows whether to bounce back or travel through. Worse still, the amount of light bouncing off changes with the thickness of the glass. This makes sense with waves because of a process called interference, but how could particles bouncing off the inside of the glass know how thick it is? QED showed that photons interact with electrons all the way through the glass, not just at the surface, so the thickness of the glass influences how many are reflected.

Photons are strange particles. Unlike particles of matter they have no inherent mass. But photons *do* have energy, and because of Einstein's famous equation $E=mc^2$, we know that energy corresponds to mass. This means that when a photon hits something, it exerts a tiny pressure.

ON THE FRONTIER

Much of our knowledge about photons comes from the ability to generate them individually. This was first done by exciting individual atoms in a process called down conversion, or using specialist lasers, but these were unreliable. More recently, tiny devices called quantum dots have been used. These are minuscule semiconductor devices fitted into a cavity between a pair of mirrors that produce individual photons when stimulated electrically.

COCKTAIL PARTY TIDBITS

❖ You may have seen scientific toys with a glass bulb containing a number of black and white paddles fixed on a spindle. In light, the paddles spin around. It's often said that this is caused by light pressure. Actually it's due to relative heating: the black sides absorb more radiant heat than the white, and it is the action of this heat on the residual air molecules in the bulb that causes the motion.

REFLECTION

When light hits a flat mirror, it bounces off at the same angle as it arrived. The closer the incoming light is to the surface of the mirror, the closer the light going out will be. When you see an image reflected in a mirror, it is as far behind the mirror as the object is from the front. The image is not really there; if you look behind the mirror as a dog or a young child might do, you won't see it. It's called a virtual image.

Things get more complex with curved mirrors. With a concave mirror, looking into the bowl of a spoon, for instance, you can get three different types of image, depending on where the object is placed. If the object is well away from the mirror the image is upside down and smaller than the object. It's in front of the mirror, so you can cast it on something, it's a real image.

If the object is closer to the mirror, depending on how close, you can either get a real image, upside down and magnified, or a virtual image—behind the mirror—the right way up and magnified. Looking into the back of our spoon, we have a convex mirror. Here the image is always virtual (behind the mirror), the right way up and smaller than the object.

Light doesn't really bounce off a mirror. A photon of light is absorbed by an electron in the mirror's surface, pushing the electron up to a higher energy level. A little later, the electron drops back down, emitting a photon that forms part of the reflection. They're not the same photons that come out as went in.

COCKTAIL PARTY TIDBITS

❖ A common puzzle is that everything in a mirror is reversed left and right, so when you lift your right hand, your mirror image lifts its left hand, but there is no similar reversal up and down. This happens because the inversion is really front and back, turning a left-handed object into a right-handed one. Imagine a rubber mask being turned inside out: if the nose of the real thing faces into the mirror, the nose of the reflection faces out of the mirror.

REFRACTION

When light passes from one transparent substance to another, it changes direction. From air into a denser substance
such as glass, it bends closer to an imaginary perpendicular
line going straight into the glass. When it comes out, the opposite happens: it bends away from the perpendicular.

The amount of bending varies with color. This isn't obvious when light passes through a glass brick, as the initial
change of direction is reversed when it comes out. However,
in a prism, a piece of glass or acrylic with a triangular cross
section, the light is bent in the same direction by both tran-
sitions, so the colors spread out and you get a spectrum or
rainbow effect.

The amount light bends depends on Snell's law. This says
that if you take the sine of the angle to the perpendicular
in the first substance and the same in the second, the ratio
of the two sines equals the ratio of the speed of light in the
two substances. This is a fixed value called the refractive index. The refractive index of a vacuum is 1, and it's around 1.5
for glass, so light travels at two thirds its maximum speed in
glass. Not sure about sines? Think of a triangle with a right
angle in one corner. Choose an angle of the triangle, and its
sine is the length of the side of the triangle opposite that
angle, divided by the length of the longest side.

Simple refraction is usually done with a flat-sided object such as a prism, but more sophisticated effects can be produced by refracting light through a curved surface, a lens.

Where light passes into a less dense substance the light bends away from the perpendicular. Imagine sending it out of a piece of glass at bigger and bigger angles to the perpendicular. Eventually it bends so far that it runs along the surface. This is the critical angle. Send light at a bigger angle to the perpendicular, and it reflects back into the substance. This effect, total internal reflection, is how light travels along a fiber-optic cable. The angle at which the light hits the inner surface of the glass fiber is so far from the perpendicular that it always gets reflected back into the fiber.

COCKTAIL PARTY TIDBITS

❖ Dutch scientist Willebrord Snell was really called Snel, but like most philosophers used a Latin name, Snellius, which seems to be the source of the misspelling.

LENSES

A lens is a shaped material—often glass or plastic—that refracts light variably, focusing rays of light at a desired point and producing magnification or clarifying an image. They have been in use since medieval times to correct eyesight, to magnify, and to focus the rays of the sun as a "burning glass."

As with curved mirrors, you can have a convex (converging) lens that bulges out in the middle, and a concave or diverging lens that gets thinner toward the middle. A converging lens can produce three different images. If the object looked at is a long way from the lens, then you end up with a real image (on the same side of the lens as the eye) that is smaller than the original and upside down. If the object is closer, it still produces an inverted real image, but this time it's magnified. Finally, if the object is very close to the lens, the image is magnified, the right way up and behind the lens (virtual); this is a magnifying glass. A diverging lens always makes an image that's virtual (on the far side of the lens), smaller than the original and the right way up.

Lenses are more flexible than mirrors because you can string them together in a row. This way, one lens can focus on the output of another, enhancing magnification.

A lens bends different colors different amounts, producing colored fringes around objects that make it difficult to see clearly; this is called chromatic aberration. The first attempts to fix this used very long telescopes, but it was gradually realized that by combining different lenses in contact with each other, for instance, a converging lens placed against a diverging one, it was possible to reduce the effect. Often, to increase the correction, the second lens is made from a different type of glass.

COCKTAIL PARTY TIDBITS

* Technically only the convex form is a lens. Originally, a lens meant something shaped like a lentil.
* Astronomers worry more about the quality of image than orientation. Astronomical telescopes generally provide an upside-down view.

PRINCIPLE OF LEAST ACTION

French mathematician Pierre de Fermat used the principle of least action to explain refraction, and Richard Feynman made it a fundamental part of quantum electrodynamics.

The principle of least action describes why a thrown basketball follows a particular route. It rises and falls along the path that keeps the difference between the ball's kinetic energy (the energy that makes it move) and potential energy (the energy that gravity gives it by pulling it downward) to a minimum. Kinetic energy increases as the ball goes faster and decreases as it slows. Potential energy goes up as the ball gets higher and reduces as it falls. The principle of least action balances the two.

The equivalent Fermat used for light is based on time, saying that light takes the quickest route. He had to make two assumptions: that light's speed isn't infinite (the speed of light wasn't known in 1661 when Fermat produced this result), and that light moves slower in a dense material such as glass than it does in air. We are used to straight lines being the quickest route between any two points, but that assumes that everything remains the same on the journey. Compare the light's journey to a lifeguard, rescuing someone drowning in the sea. The obvious route is to head straight for the drowning person. But the lifeguard can run much faster on the beach than in the water. By taking a longer path on the sand, then bending inward and taking a more direct path in the water, the lifeguard gets there quicker. Similarly, a light ray can reduce its journey time by spending longer in air and less time in glass. The angle that minimizes the journey time is the one that occurs.

ON THE FRONTIER

When Richard Feynman was inspired by the principle of least action to come up with quantum electrodynamics, he imagined not just the "best" path but every single possible path a particle could take in getting from A to B. QED takes a sum of every single possible path combined with the probability of that path occurring to predict a particle's behavior.

COCKTAIL PARTY TIDBITS

❖ The analogy with a lifeguard has led to Fermat's approach being called the *Baywatch* principle.

THE SPEED OF LIGHT

Light travels at around 186,000 miles per second in a vacuum, taking a little over eight minutes to reach the Earth from the Sun.

For a long time there was a debate over how fast light traveled. Some thought it instantaneous. Descartes, for instance, thought seeing light from a distance was like being poked by a very long pool cue. The moment the source gives off the light, he thought, was like someone giving one end of the cue a push: the other end of the invisible cue immediately pushes against the eye, registering sight.

Galileo tried measuring the speed of light by sending a servant up a hill and timing lantern flashes, but he discovered that human response time was much slower than light itself. It was only when Danish astronomer Ole Roemer was making measurements of the movement of Jupiter's moons that he discovered the changes in relative position of the Earth and Jupiter were big enough for the different times the light took to arrive to have a visible effect on the moons' motion, allowing him to estimate the speed.

Since Roemer, the speed of light was measured with a battery of mechanical devices, often involving fast rotating mirrors and cog wheels, before electronic measuring devices enabled us to pin down the speed of light with modern accuracy.

Light can exist at only one speed. But normally, if you move alongside something else at the same speed, it doesn't move with respect to you. Einstein realized if he rode alongside a sunbeam, it should stop and cease to exist—but this doesn't happen. Bizarrely, however you move relative to it, he deduced that light always goes at the same speed. It was this realization that made all the strange physics of relativity possible.

COCKTAIL PARTY TIDBITS

❖ In 1983, the speed of light was fixed at 299,792,458 meters per second. It will never change, because the meter has since been defined as 1/299,792,458th of the distance light travels in a second. It's a shame they didn't make it 1/300,000,000th.

THE FIRST TELESCOPES

It's often said that Galileo invented the telescope. He didn't, but his first telescope was a success because of crafty political maneuvering. Galileo heard that a Dutchman was coming to Venice with a telescope. A friend of Galileo's was asked to investigate the new technology. Galileo got his friend to keep the Dutch inventor busy, giving himself time to build his own telescope and get to Venice first.

Early telescopes were refracting devices, using a large lens to collect the light and a second to magnify the image produced by the first. But simple lenses cause distortion by bending different colors by different amounts. Trying to solve this problem produced some bizarre devices. It had been noticed that the shorter the focal length (the distance over which the lens focuses the rays of light) the worse the distortions. So Galileo's successors went for length. One of the best-known examples, made by the Polish astronomer Hervelius, was 45 meters (150 feet) long.

ON THE FRONTIER

Modern research suggests that the real inventor of the telescope may be the English father and son Leonard and Thomas Digges. Leonard was an adventurer who had the considerable luck to survive a failed revolution against Queen Mary, and Thomas was a noted scholar. When Thomas wrote about his father's work after Leonard's death, he claimed that they had used "perspective glasses" to see distant objects. Queen Elizabeth's court asked William Bourne, an expert on military technology, to describe the telescope and its limitations (it had a very narrow field of view), suggesting that a device was built.

COCKTAIL PARTY TIDBITS

❖ English friar Roger Bacon may have built a telescope in the thirteenth century. Bacon wrote, "Lenses are contrived so that the most distant objects appear near at hand and vice versa. . . . We may read the smallest letters at an incredible distance, we may see objects however small they may be, and we may cause the stars to appear wherever we wish."

TELESCOPES COME OF AGE

Newton and others worked out that using a mirror to collect light, rather than a lens, would prevent the distortions of refraction. These reflecting telescopes became the preferred astronomical instruments.

The first of the really big reflectors with a 49-inch mirror was built by William Herschel in Slough, England, in the early 1800s, but this proved difficult to keep stable on its huge steerable wooden framework. In principle it was bettered by another instrument with a 72-inch mirror. Built in Ireland in 1845, this "Leviathan of Parsonstown" overcame the stability problem by being pivoted between two brick walls. This limited how much sky it could traverse, but kept it solid. Unfortunately the weather was so bad that it could rarely be used.

It was only when American observatories took the lead that really great telescopes could be built. With drier, clearer weather, observatories like Mount Wilson and Mount Palomar achieved amazing results with huge reflectors containing 100-inch (1917) and 200-inch (1948) mirrors, respectively.

ON THE FRONTIER

Recent telescopes rely on computer technology to get around the limitations of their predecessors. One approach is to split the mirror into segments. These weigh a lot less than a single piece of glass and so are easier to manufacture and to wrangle. Computers knit together the imagery from the mirror segments. Other telescopes use adaptive optics. These make rapid movements of mirrors, or distort the mirror surface, to undo distortions caused by vibration and air currents.

COCKTAIL PARTY TIDBITS

❖ The Mount Palomar telescope took 15 years to build, interrupted by World War II. The mirror was made from 65 tons of glass.

❖ At the time of writing, the world's biggest telescope is the Gran Telescopio on the Canary Islands, with a segmented mirror 10.4 meters (409 inches) across. Close behind are the two Keck telescopes on Mauna Kea, Hawaii, at 10 meters (394 inches) each.

❖ The Keck telescopes, acting together, could distinguish car headlights 16,000 miles away.

TELESCOPES REACHING FARTHER

The most famous telescope today is probably the Hubble Space Telescope. This was launched on April 25, 1990, from the shuttle *Discovery*, only to find that there was a fatal flaw in the shape of the mirror. A repair mission was undertaken between December 2 and 13, 1993, after which the telescope functioned perfectly. Although the Hubble mirror is a relatively puny 94 inches, being outside the Earth's atmosphere means that it can observe with unparalleled clarity.

An alternative route to getting around the limitations of visible light is to use other parts of the electromagnetic spectrum. Practically the entire spectrum is used, from high-energy gamma rays and X-rays down to radio. Each part of the spectrum enables different types of observation, often complementing the visual. Best known are the radio telescopes.

Back in the 1930s and 1940s, engineers working on radio receivers accidentally discovered that there were radio sources out in the universe, the Sun, for example. Radio telescopes are much less precise than their optical equivalents, but they can pull in signals across a much wider area. When the biggest optical telescope was 200 inches, radio telescopes were built at 250 feet across. These can be steered around the sky, but the biggest single radio telescope at Aricebo in Puerto Rico is a fixed dish, 1,000 feet across. Modern radio telescopes use an array of dishes to provide the equivalent of a vast antenna.

ON THE FRONTIER

There are plenty of other satellite-based telescopes, often dealing with different parts of the electromagnetic spectrum, but the successor to the Hubble will be the James Webb Space Telescope. This will detect both visible and infrared light, allowing the telescope to see farther than using visible light alone. The Webb mirror, made up of 18 segments coated in 24 karat gold, will be 6.5 meters (256 inches) across. The Webb, expected to launch in 2013, is named after James E. Webb, the second of NASA's administrators.

COCKTAIL PARTY TIDBITS

❖ The fault on the Hubble telescope was that the mirror was too shallow. The total error was just one fiftieth of the width of a human hair.

MICROSCOPES

Magnifying glasses were not uncommon from medieval times, and magnifying mirrors (such as a shaving mirror) had been around since the Ancient Greeks.

The simple action of putting two lenses together in a tube transformed our ability to delve into the realities of microscopic life. In a two-lens microscope, a lens close to the object being studied produces a magnified image on the opposite side of the lens. The second lens, the eyepiece, then acts as a magnifying glass, focused on this enlarged image.

Compound microscopes were first developed by Dutch lens grinders Hans and Zacharias Janssen. When the first device was assembled, around 1590, Zacharias was only a boy, so his father Hans probably deserves the bulk of the credit. Another name frequently connected with early microscopes is Anton van Leeuwenhoek. He was responsible for one of the first breakthroughs using a microscope, discovering bacteria in 1674, but his instrument was only a single lens, so was little more than a powerful magnifying glass.

ON THE FRONTIER

Optical microscopes are no longer the only way to observe the very small. Electrons, as quantum particles, can be used in place of photons of light to examine an object, but with their electrical charge and slower speed, electrons are much easier to manipulate, proving ideal as a way of scanning the surface of an object and displaying its detailed texture. An electron microscope can magnify around 1,000 times more than the best optical microscope. The first was built in 1931 by German scientists Ernst Ruska and Max Knoll.

COCKTAIL PARTY TIDBITS

❖ Scanning tunneling electron microscopes can also manipulate very small things. In 1989 Don Eigler of the Almaden Research Center used a scanning tunneling microscope to spell out the letters IBM using individual xenon atoms on a nickel surface.

QUANTUM LENSES

THE BASICS

The idea of using a lens to focus light is so firmly fixed in our minds that it can be easy to forget that focusing is nothing more than a change in direction of photons. With our understanding of the quantum nature of light, optical devices far exceeding the capabilities of lenses can now be made.

The two principal approaches are metamaterials and photonic crystals, each modifying the way light behaves by interacting with individual photons. Metamaterials consist of layers of lattices or a pattern of tiny holes in a metallic sheet; this structure gives them their optical properties. Photonic crystals act on light as semiconductors act on electricity, providing unparalleled precision of control.

Metamaterials go beyond anything found in nature. Natural transparent materials have a positive refractive index. When light hits a block of glass, it bends in toward the perpendicular. However, when light hits a metamaterial, it bends in the opposite direction. The metamaterial has a negative refractive index. This means that metamaterials can manipulate light in unexpected ways.

Unlike metamaterials, photonic lattices do occur in nature, although only in imperfect forms. Both the swirly glittering appearance of an opal and the iridescence of a peacock's tail are caused by photonic lattices. But artificially created photonic crystals can do much more than produce a pretty effect. Photonic lattices are essential components if we are to build optical computers using light rather than electricity to convey signals inside the computer.

Conventional microscopes can't focus on anything smaller than the wavelength of light used. But this limitation is shattered by metamaterial superlenses. Not only can such metamaterial lenses be built for a fraction of the cost of an electron microscope, they enable a different kind of observation, just as radio telescopes and visible light telescopes work together in astronomy.

COCKTAIL PARTY TIDBITS

❖ Like a Harry Potter invisibility cloak, metamaterials can make objects disappear. Because of their negative refractive index, metamaterials bend light around an object. This has already been done on a small scale with microwaves, but is harder with visible light, where the material absorbs much of the light. However, there are alternative mechanisms that can optically amplify the output of the metamaterial, so we may still have invisibility cloaking in the not too distant future.

POLARIZATION

In 1669, Scandinavian experimenter Erasmus Bartholin believed he had found two different types of light. When he put a block of a quartz called Iceland spar on top of a piece of paper with a straight line drawn on it, he saw not one but two lines. It was as if there were two kinds of light, one bent more than the other by the crystal, producing two, clear, separated images.

The significance of this discovery was not apparent until Augustin Fresnel, a French military road builder, used it to help understand light. Fresnel realized that if light were a wave like a ripple in a piece of string, it could move up and down, or side to side. Fresnel imagined that Iceland spar contained an invisible grid of slots, some horizontal, some vertical. The side-to-side light would pass through the horizontal slots, and the up-and-down light through the vertical. The result would be to split the two types of light.

As far as our eyes are concerned, there's no difference between light waves whichever direction they ripple. The Sun's light is a jumble of rays oriented in every possible direction, leaving some to pass through each of the Iceland spar's grids. The orientation of a light ray's ripple is its polarization.

There are more complex polarizations called circular and elliptical polarization that rotate as the wave moves along. When taking the photon view of light, each photon has a polarization: it is a quantum property of the photon (like its spin) with an associated direction.

❖ American student Edwin Land became fascinated by polarized light while at Harvard in 1926. It was known that reflected light was partly polarized. Land felt that the polarization effect had commercial value. Still only 18, he took a leave of absence from Harvard. The result of his experiments was Polaroid, a plastic sheet with polarizing crystals embedded in it. By 1937, his garage laboratory had become the Polaroid Corporation. The glare that irritates motorists or ruins photographs can be cut down dramatically with a piece of Polaroid material in the right orientation.

REDSHIFT

Redshift is light's equivalent of the Doppler effect, but instead of sound shifting in pitch, the light shifts in color. Light is bluer if the object emitting it is moving toward us (a blueshift), and more red if it's moving away (redshift). Imagine a light wave flowing out from the object. Before the next ripple can come out, the object will be a bit closer than it was a moment ago, so the effect is that the wave is squashed up (that means shorter wavelength and higher frequency); it is moved toward the blue. It undergoes a blueshift.

If you prefer the photon view of light, a blueshift is an increase in the energy of the photons. The movement of the emitting body toward us gives the photons a boost of energy, just as a baseball thrown toward us from a moving car hits us with more energy than one thrown by a stationary pitcher. As light can't speed up, that increase in energy produces a shift up the energy spectrum toward the blue. Moving away has the opposite effect, giving a redshift.

ON THE FRONTIER

When we look at distant objects in the night sky we can tell how they are moving with respect to us because they are blueshifted or redshifted. Almost all galaxies are redshifted, the first evidence that the universe was expanding. But to know this, it had to be possible to measure the color shift.

On the way out of a star, light passes through the star's outer layers. As it does, some of the frequencies present are absorbed, leaving a series of black lines in the color spectrum. Each element produces characteristic black lines, and from these it's possible to deduce the elements in the star. These lines are detected using a spectroscope, a device which at its simplest is just a prism, splitting apart the colors in the light, and a microscope.

These characteristic black lines always appear in recognizable patterns, but with a redshift, the patterns are moved toward the red end of the spectrum.

COCKTAIL PARTY TIDBITS

❖ **Not every galaxy has redshift: the nearest to our own, the Andromeda galaxy, is blueshifted. This is because gravity is pulling the closer galaxies together faster than the universe is expanding. Eventually our galaxy will collide with the Andromeda galaxy, but not for billions of years.**

LASERS

The laser's initial form was an accidental discovery by Russian scientists Nikolai Basov and Alexander Prochorov, investigating the behavior of the pungent gas ammonia in 1954. Basov and Prochorov found that light in the nonvisible microwave region triggered the release of photons from ammonia. Generated in a sealed chamber, those photons could themselves stimulate further photons, a Ponzi scheme approach to producing light. Because of the way the light was stimulated, the photons were all in phase. They described it as Microwave Amplification by the Stimulated Emission of Radiation, a maser for short.

By 1960 the American Theodore Harold Maiman developed an equivalent device for visible light. The concept was the subject of a patent battle between American physicist Arthur Leonard Schawlow and another American, Gordon Gould. Gould was eventually recognized as the theoretical originator of the visible maser that Maiman built. Gould called his concept the laser, replacing "microwave" in maser with "light."

Maiman's device used a ruby to produce the stimulated emission, giving out a deep red light. The light was stimulated using a flash tube like a huge photographic flash unit. Inside the ruby, the light passed backward and forward, hitting mirrors at either end, stimulating more photons on each pass. One mirror was partially silvered, allowing some photons to escape.

Because of the way that laser light is produced it is entirely different from the rays of the Sun or an incandescent bulb. The phase of each photon is synchronized. The result is a powerful beam of light of a single color that is not easily scattered and dispersed as ordinary light is.

ON THE FRONTIER

In 1917 Einstein predicted that it would be possible to set off a chain reaction producing light, which he described as stimulated emission. According to Einstein's theory, an electron in an atom can be pushed into a high-energy state when it is hit by a photon, leaving it like a bucket of water sitting over an open door. Another photon, hitting that electron, would both be re-emitted and would trigger the electron to release a second photon, as if the bucket were knocked off the door by a stream of water from a hose.

COCKTAIL PARTY TIDBITS

❖ A laser beam can be bounced off the moon and returned as a tight ray.

HOLOGRAMS

Soon after the Second World War, Hungarian-born British scientist Dennis Gabor was thinking about the way we see objects. Imagine looking through a glass window at a mug on a table. Stand to the left and you see a certain view of the mug, perhaps the handle and the front side. Move around to the right and the view changes. All the light required to make up these different views is falling on the window glass. So if there were a way to take a snapshot of every ray of light traveling from the mug to the glass, you should be able to re-create the view from the window.

To cope with all the photons coming from different directions you would need to distinguish not just how bright a particular point is, as an ordinary photograph does, but also the phase of each photon. To do this, Gabor imagined using a second beam of light directed onto the glass. The two light sources would interfere with each other as do the beams passing through Young's slits. The resultant pattern would indicate the phase of each photon when it hit the glass.

Gabor couldn't make one of these pictures (they were called "holograms" from the Greek *holos* meaning whole and *grapho* to write), because they would only work with a special kind of light source that didn't exist, one with all the photons in phase.

Any part of a hologram includes light coming from many different directions. Just as looking through half a window you still see the whole view, so breaking a hologram in half still shows you the whole picture.

COCKTAIL PARTY TIDBITS

❖ It took only a couple of years after the laser was invented in 1960 before Emmett Leith and Juris Upatnieks at the University of Michigan produced the first true hologram, a still life of a model train and a pair of stuffed pigeons.

❖ The security "holograms" on banknotes and credit cards are not true holograms. These images have two or three layers using optical technology to give the appearance of depth.

STOPPING LIGHT

U.S. researchers have brought light to a standstill with special, low-temperature materials. In 1998, Lene Vestergaad Hau and her team at Edwin Land's Rowland Institute for Science at Harvard University set up an experiment where two lasers were blasted through a vessel containing sodium atoms cooled to form that special state of matter called a Bose–Einstein condensate.

Normally the condensate would be opaque, but the first laser creates a sort of ladder through the condensate that the second light beam can claw its way along, at vastly reduced speeds. Within a year, Hau's team had pushed down the speed to below a meter per second.

More recently, Hau's team has discovered that if the first "coupling" laser is gradually decreased in power until it is switched off, the second beam is swallowed up in the material. The result is a strange mix of matter and light called a "dark state." The trapped light only comes out again when the coupling laser is reestablished. Although the light is stuck inside the dark state, it has to move to exist. It is like a pacing animal in a cage, always moving but never getting out.

COCKTAIL PARTY TIDBITS

❖ Science fiction author Bob Shaw created the concept of slow glass, a material that took light years to get through. With such remarkable glass it would only take a site in front of a beautiful view to create stunning windows. If light takes a year to get from one side of the glass to the other, then one year after the windowpane is put in position, the first glimpse of the landscape will reach the other side. The glass can then be moved to another location, taking the view with it.

❖ When a TV crew came to film in Hau's laboratory they were disappointed as the lasers in use were invisible, so they set up a smoke machine. The result was a total collapse of the experiment, which had to be shut down for days while the air cleared.

FASTER THAN LIGHT

Special relativity says that nothing travels faster than light, but this doesn't allow for tunneling. This enables a quantum particle to go from one side of a barrier to another without traveling through the space in between. So if, for instance, a photon went across a meter of space, then through a meter-wide barrier, then across another meter of space, it would have covered three meters in the time light takes to cover two; it travels 1.5 times the speed of light.

These barriers can be made in several different ways, using special versions of metal tubes that carry microwaves called waveguides by using photonic lattices, or by placing two prisms close to each other, where a totally internally reflected beam tunnels across the gap. However, in all cases, the distance is so short that the immense speed can't be used. By the time you measure any information, you have already lost the time advantage. And the thicker the barrier, the fewer photons get through. Make it thick enough to be able to do something in the timescale and nothing gets through.

ON THE FRONTIER

Moving at this remarkable rate, the experimenters' pulses of light slip back against the time stream like salmon fighting their way up a river. If a usable signal traveled along such a faster-than-light beam it would arrive before it was sent. Although at first sight this has positive possibilities, like checking out lottery results ahead of time, it would destroy the whole basis of causality.

COCKTAIL PARTY TIDBITS

❖ When Günter Nimtz of the University of Cologne in Germany documented this effect, he was challenged by Raymond Chiao of the University of California at Berkeley. Chiao believed it was impossible to send a signal through the tunneling barrier. Nimtz responded by sending a recording of Mozart's 40th Symphony at four times the speed of light.

FOUR

RELATIVITY

GALILEO'S RELATIVITY

Ask most people who came up with relativity and they'll say "Einstein." They're wrong. It was Galileo. He imagined being on a ship on a totally smooth sea, moving steadily, not getting slower or quicker. If you were in a cabin without windows, how could you tell you were moving? Everything in the cabin would move the same way. You wouldn't feel the motion; you only do that when you are accelerating or decelerating. Water in a fish tank would not slop about. There would be no indication of movement.

Galileo said that any two observers moving at constant speed and direction with respect to each other will obtain the same results for all mechanical experiments. Provided two experiments continued moving smoothly, there wouldn't be any way to tell them apart. There are some other things that need to be fixed here—they both need to be under the same gravity, air pressure, and so on—but the principle is sound.

ON THE FRONTIER

At the heart of Galileo's observation is that there is no such thing as absolute motion. All movement has to be with respect to something else. If, for instance, two trains move alongside each other on parallel lines at the same speed, the second train is not moving as far as the first train is concerned (and vice versa). That second train doesn't just *seem* to be stopped from the first train, it *is* stopped relative to the train. Similarly, when you stand still, you are stopped relative to the Earth. But the Earth is spinning on its axis, flying around the Sun, and flashing through space with our galaxy.

COCKTAIL PARTY TIDBITS

❖ Galileo wasn't the first in his family to challenge authority. His father Vincenzio, a court musician, wrote a book on music in which he wrote that those who rely purely on authority (as happened in science then) "act very absurdly."

SPECIAL RELATIVITY

Like Galileo, Albert Einstein was aware that there was no fixed reference point compared to which everything is moving or is static. We can only describe an object's velocity relative to something else. But this caused a problem for light.

Einstein imagined riding along on a sunbeam. If he had been able to do that, then the light beam would stop. But Einstein knew that the Scottish scientist James Clerk Maxwell had shown that light was an interplay of magnetism and electricity that could be self-sustaining at only one particular speed. As light plainly did exist, Einstein came to the startling conclusion that light uniquely always moves at the same speed, around 300,000 kilometers per second in a vacuum. And from this simple conclusion all the strange predictions of special relativity emerged.

ON THE FRONTIER

Plugging a fixed speed for light into Galileo and Newton's equations of motion produced bizarre results. As a body gets closer to the speed of light it contracts in size, gets more massive, and the passage of time on the body gets slower. All these effects are seen from the place the body moves relative to. If the moving object is a spaceship, for instance, people on that ship don't see the distortion, because they are moving at the same speed. If anything could reach the speed of light its mass would become infinite; it would be infinitely small and time would stop. These changes are not hypothetical. For example, we see fast-moving particles called mesons in the atmosphere, which should have decayed long before they reached the Earth's surface. They get through because they move so fast relative to the Earth that time slows for them.

COCKTAIL PARTY TIDBITS

❖ Albert Einstein said, "When a man sits with a pretty girl for an hour, it seems like a minute. But let him sit on a hot stove for a minute— and it's longer than any hour. That's relativity." This was the abstract of a paper he allegedly published in a journal called *Journal of Exothermic Science and Technology.* This is usually portrayed as a genuine, if humorous, academic contribution, but the initials of the journal suggest that Einstein made the whole thing up.

❖ When Einstein devised special relativity he was not a professional scientist. He was an administrator at the patent office in Bern, Switzerland.

THE TWINS PARADOX

It's hard to imagine anything more unlikely than twins with different ages, but that's one of the stranger consequences of relativity.

Let's imagine identical twins Donna and Gill, in a tearful separation. Donna is seeing off Gill on the first flight to the stars on their 21st birthday. Gill's ship shoots off at 99 percent of the speed of light. Seven years later, Donna is celebrating her 28th birthday, but if she could see Gill, there would only be 22 candles on Gill's cake. This same effect takes place on the reverse journey. So when Gill returns, having spent two years on a distant planet, Donna is 37, but Gill is just 25.

There is an apparent flaw in this science fiction scenario. Relativity says there is no difference between saying the spaceship flies away from the Earth, and saying the Earth flies away from the spaceship. To the people on Gill's spaceship *they* are stationary and the Earth flies away from them. So from Gill's viewpoint it should be Donna who is younger. But the twin paradox really works, because the situation isn't really symmetrical.

It's true that each twin would think the other was aging slower on the journey. But at the endpoint, there is a difference between the Earth and the spaceship. A force is applied to the ship to slow it down, turn it around, and send it back toward the Earth. The Earth doesn't undergo this acceleration. So Gill goes through a different relativistic process than Donna, and it is Gill who arrives home younger.

ON THE FRONTIER

A similar effect has been demonstrated for real using atomic clocks. A pair of atomic clocks are synchronized, then one is flown around the Earth, while the other stays in place. The clock that had the acceleration (going around in a circle implies accelerating, as acceleration is a change in velocity, which is speed and direction) was slow compared to the "stationary" clock.

COCKTAIL PARTY TIDBITS

❖ When the crew of the Salyut space station landed in 1988 after a year in orbit they were about one hundredth of a second younger than if they'd stayed at home.

SIMULTANEITY

It's easy enough to say whether two events at the same spot happen simultaneously. But what if two bolts of lightning strike buildings in the same street, five blocks apart? How could I tell if the lightning strikes were simultaneous? I could have a clock at each location and compare timings, but how could I be sure the clocks were synchronized?

The only way to be certain would be to stand halfway between the two strikes. If the light from the lightning bolts reached me in the middle at the same time, they would be simultaneous. That's fine, but what if I made my observation from a bus heading west up the street? In the time it took the lightning flashes to reach me, I would have moved. So the flash from the west end of the street would have reached me before the flash from the east end. For me, the lightning strikes wouldn't be simultaneous; the west one would have happened first.

Because relativity doesn't allow me to define any particular way of moving steadily as special, my view from the bus is as valid as any other. So simultaneity is relative: it depends on how I move with respect to the events.

The relativity of simultaneity is illustrated by the ladder paradox. Special relativity tells us that an object moving with respect to us is smaller than that object at rest. Imagine you had a ladder that's too long to fit into a garage. Move the ladder fast enough into the garage, and the ladder would be contracted enough (from the garage's viewpoint) to fit inside. But from the ladder's viewpoint the garage shrank, and the ladder won't fit. Let's imagine we've very quick doors on both ends of the garage. From the viewpoint of the garage, we can briefly close both doors simultaneously as the ladder shoots through the garage. But from the ladder's viewpoint the doors won't close simultaneously; the far door shuts while the ladder is on the way in, and the near door as the ladder emerges.

COCKTAIL PARTY TIDBITS

❖ Physicists use Minkowski diagrams to show the relativity of simultaneity. These should be four-dimensional charts (three dimensions of space plus one of time), but to make them more practical they are usually shown flat with the time dimension running up the page and a space dimension running side to side.

TIME TRAVEL

Traveling faster than light implies traveling back in time, making your arrival time earlier than your departure time. Once this happens the whole flow of reality is at risk, as dire paradoxes of time travel come flooding in.

Imagine a radio transmitter/receiver that could send a message a fraction of a second through time. Let's assume this radio can be switched on or off by radio control. We use the radio to send a message back a few moments before to switch itself off. So it is off when we send the message. So the message isn't sent. So it's still on . . . It's a loop of logical impossibility. This strangeness can be dressed up more dramatically as the grandfather paradox. What if someone went back in time and killed her own grandfather before he met her grandmother? The killer wouldn't be born, so she couldn't have gone back in the first place.

This is so unsettling that some physicists impose a "causal ordering postulate" saying that two events that are causally connected (one triggers the other) must always come in the same order. This is no physical law, though: it's more like an academic raised eyebrow.

Time travel is possible. Photons can be sent faster than light through a tunneling barrier, but the time shift is too short to use. Quantum entanglement communicates instantly at any distance, but can't send a message. Among hypothetical time machines are near-infinite rotating cylinders of neutron star material and wormholes in space, but neither is physically practical. The best hope comes from physicist Ronald Mallett, who has spent a working life looking for time travel mechanisms to communicate with his father, who died when Mallett was a child. He believes that laser beams traveling around a ring of mirrors can produce an effect called frame dragging which could enable time travel into the past.

COCKTAIL PARTY TIDBITS

❖ When the telegraph was first introduced in the nineteenth century it allowed a kind of time travel. It was common practice to bet on a horse race up to the time a messenger arrived with the result. The telegraph allowed sneaky sportsmen to get around this limit. In one example, a telegraph message was sent from a race course to London, saying, "Your luggage and tartan will be safe by the next train." Tartan was a codeword identifying the colors of the winning jockey, allowing a bet to be placed before the result was officially known.

WHERE ARE THE TIME TRAVELERS?

Physicist Stephen Hawking has asked why we haven't seen any time travelers if time travel is possible, but there are plenty of reasons why this could be the case. If a civilization had the technology to travel through time, they would also be able to conceal their existence from us. And many of the ideas for time travel will only work on light, rather than a physical person.

To make it even less surprising that we haven't had any contact from the future, most time machines won't work before the device was first built. For example, imagine we've invented a communicator that works instantly across any distance. We send a probe out into space for 20 years at half the speed of light. Then, for us, time will have slowed down on the probe. A message we send out in 2020 would arrive $5\frac{3}{4}$ years earlier.

But from the probe's viewpoint, the Earth moved away from it at half the speed of light. It would see the Earth's clock as $5\frac{3}{4}$ years behind its own. So sending that message instantly back to the Earth would mean it arrived around $11\frac{1}{2}$ years before it was first sent. In practice, we don't have a practical way to send a signal faster than light. But if we could, this device would send a message back 11 years, after the probe had been flying for 20 years. However fast it went, it couldn't get a message back before the probe was launched.

ON THE FRONTIER

On May 7, 2005, MIT ran a convention for time travelers. The idea was that people in the future would read about the convention—in this book, for instance—and travel back to take part. Sadly, no certified time travelers turned up. Similarly, there is a plaque in Perth, Australia, asking time travelers to arrive on that spot on March 31, 2005. Once more, no one came.

COCKTAIL PARTY TIDBITS

❖ On *Star Trek*, the crew occasionally travels in time, for instance using a slingshot maneuver around the Sun. In practice they needn't bother with fancy techniques. Every time they use the warp drive and fly faster than light, they also travel backward in time, but this is ignored to keep the show simple.

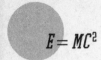

$E = MC^2$

The most famous equation in science is Einstein's $E=mc^2$. We know that special relativity means that objects get heavier as they get closer to the speed of light. But where does that mass come from? It comes from the energy accelerating the object. At normal speeds, the more energy we put in, the faster the object goes: kinetic energy, the energy of movement, is simply $\frac{1}{2} mv^2$ where m is the mass and v is the velocity. But as the object gets closer to the speed of light, most of the energy goes into mass rather than acceleration. The increases in velocity become smaller and smaller as the mass increases.

We can also see energy being converted into entirely new particles. When a high-energy particle collider smashes particles into each other at near the speed of light, the energy of the collision is converted into particles that spring into existence out of nowhere.

Something that's more familiar, though, is the reverse process: the conversion of mass into energy. On an incredibly small scale, this happens every time we burn something. The chemical bonds that link atoms are broken in this process, resulting in a tiny loss in mass. But much more dramatic is when two or more nuclear particles fuse together, producing a new nucleus with lower mass. We can see this happening in the sky all the time; it's the process that powers the Sun. It used to be thought that the Sun was on fire. But in the nineteenth century it was realized that if the Sun were made of coal it would last only a few thousand years, and it was already known that the Earth was much older. It was only the discovery of atomic fusion that overcame the problem.

ON THE FRONTIER

Nuclear power stations use nuclear fission, powered by split atomic nuclei. If we could duplicate the mechanism of the Sun, nuclear fusion, we could produce cleaner, safer nuclear energy. Most estimates put it at 30 to 50 years away from being practical.

COCKTAIL PARTY TIDBITS

❖ Whenever we put energy into something, we increase its mass. A hot cup of coffee weighs more than a cold cup.

EQUIVALENCE

Einstein has described what brought him to his masterpiece, general relativity: "The breakthrough came suddenly one day. I was sitting in a chair in my patent office in Bern. Suddenly the thought struck me: If a man falls freely, he does not feel his own weight. I was taken aback. This simple thought experiment made a deep impression on me. This led me to the theory of gravity."

Einstein's insight was later called the principle of equivalence: the effects of gravity and acceleration are indistinguishable. This led Einstein to devise a theory that both explained the way gravity worked and generalized special relativity to deal with accelerating bodies.

Useful though it was, the principle of equivalence is technically wrong. It states that if someone is in a spaceship without windows, accelerating at 32 feet per second per second, they couldn't distinguish this from the ship being on the surface of the Earth with an equivalent acceleration due to gravity. In reality, however, there *is* a distinction. The whole spaceship accelerates at the same rate, so an experiment undertaken near the top of the ship will have exactly the same result as an experiment at the bottom. But gravity decreases as you get farther from the Earth. So an experiment in the top of the stationary ship on the Earth's surface would have subtly different results from an experiment at its bottom.

COCKTAIL PARTY TIDBITS

- ❖ Equivalence predicts that gravity bends light. Imagine a beam of light passing across the inside of our accelerating spaceship. In the time it takes to cross the ship, the ship will have moved, so the light will hit the opposite wall slightly lower than expected. The acceleration bends the light beam. As equivalence says whatever is observed under acceleration is also observed under gravity, a light beam crossing the ship sitting on the Earth will be bent downward by the Earth's gravitational pull.

GENERAL RELATIVITY

Einstein didn't just believe that gravity and acceleration acted similarly. He thought that they were the same thing. Gravity isn't a regular force such as the force you use to pull a car along. With a "real" force, the acceleration you get depends on the mass of the thing you are moving. But gravity has the same acceleration whatever the object's mass.

But if there isn't really a force due to gravity, then the Earth should travel along in a straight line. What Einstein realized is that effectively it does. But a body with mass makes space warp. So the "straight line" that the Earth follows curves around the Sun.

It's difficult to imagine space warping because we can only envisage three dimensions, and it's hard to envisage all three warping simultaneously. The usual approach is to think of a vast rubber sheet. This is a two-dimensional model of space; we're just ignoring the third dimension. If we place a heavy object such as a bowling ball on the sheet it distorts into another dimension. A smaller object placed nearby will roll down the slope until it hits the bowling ball. This is a useful picture for how space is curved by objects with mass, accelerating other objects toward them without truly exerting a force.

The experiment in a spaceship where a beam of light bends downward demonstrates the general relativity picture of curved space. Light takes the most direct path. If space weren't curved, this would be a straight line, but if space were distorted, the most direct route would be a curve.

❖ General relativity warps a four-dimensional environment, spacetime, with time as the fourth dimension. When an object is subject to gravity, the bending of spacetime means that time flows more slowly relative to somewhere that isn't subject to gravity. The heavier the gravitational pull is, the more time slows.

FORCE

A force makes things happen. Before we look at the fundamental forces in the universe, it's worth thinking for a moment about that most typical outcome of force, movement. The result is described by three simple rules: Newton's laws of motion.

The first law seems facile. It says that a body will stay the way it is (still or moving at a constant speed) unless a force changes things. It seems a bit "so what?" but actually it's profound. Without this, none of the more complex aspects of forces would make sense.

The second law tells us how much a force changes things. The force is the same as the mass of the object times the acceleration it receives. The bigger the force, for a given mass, the more acceleration there is. Apply the same force to two objects with different masses and the one with less mass accelerates more.

The third law says that if one body exerts a force on a second body, then the second body exerts an equal force on the first body, in the opposite direction. So, for example, when your car accelerates, a force acts on it to move it forward. At the same time, your car exerts the same size force on the Earth in the opposite direction. But because the Earth has a lot more mass than your car, the acceleration the Earth gets is tiny.

ON THE FRONTIER

We now know that Newton's laws are an approximation. Close to the speed of light, you have to include relativistic factors, but at everyday speeds you end up with Newton's laws.

COCKTAIL PARTY TIDBITS

❖ The word "law" here is confusing. Science works by guessing at how the world works and from this "model," predicting what will happen. Newton's laws are based on a model that makes good, but not perfect predictions. In honor of Newton, force is measured in "newtons."

FIELDS OR PARTICLES?

Back in the 1800s, when Michael Faraday was experiment-
ing with electricity and magnetism, he devised the concept
of a field. This was the area of influence of a force, such as
electricity or magnetism. (It was later discovered that electricity and magnetism were two sides of the same coin.) The
field was like an invisible weather map of the strength of the
force. As an object moved through the field it cut through
lines of force; the closer together these were, the stronger
the force was.

The modern Standard Model of forces uses particles called
bosons to transmit forces from place to place. Fields are an
alternative model that can be helpful in the understanding
of forces, and it's a model that is still extremely valuable.

Underlying every action is one or more of the four fundamental forces. These are gravity, electromagnetism, weak
nuclear force, and strong nuclear force.

Quantum physics speculates there is a field called the
Higgs field, the equivalent of transmitting a particle called
a Higgs boson. This gives all the particles their property of

mass. (Technically there are fields equivalent to each of the boson particles in the Standard Model.)

One of the interpretations of quantum physics—David Bohm's quantum potential interpretation—adds another field called the psi field. This permeates space from each quantum particle, enabling the particles to interact as if distance didn't exist, causing some of the stranger tricks of quantum particles, but Bohm's interpretation is not widely supported.

COCKTAIL PARTY TIDBITS

❖ Fields can give rise to problems for particles like electrons, which are considered to be points with no size. The field lines radiate out from the point like spokes on a wheel, but this means that the force should head off to infinity as you approach the electron.

ACTION AT A DISTANCE

If we want to act on something that isn't directly connected to us, we need to get something from us to the object on which we wish to act. Often this "something" involves direct contact: I reach over and pick up my coffee cup to get it moving toward my mouth. But if we want to act without crossing the gap that separates us from the object, we need to send an intermediary.

Imagine that you want to knock a can off a fence a few yards away. You can't just look at the can and make it jump into the air; you have to throw a stone at it. Your hand pushes the stone; the stone travels through the air and hits the can. As long as your aim is good (and the can isn't wedged in place), the can falls off.

Similarly, if I want to speak to someone on the other side of a room, my vocal chords vibrate, pushing against the nearest air molecules. These send a train of sound waves, rippling molecules across the gap, until those vibrations get to the other person's ear, start her eardrum vibrating and result in my voice being heard. In the first case, the ball was the intermediary, in the second the sound wave, but in both cases *something* traveled from A to B. In the case of the four fundamental forces, it is the particles called bosons that provide this "something" that goes from A to B and enables one particle of matter to influence another.

The only true action at a distance is quantum entanglement. This quantum phenomenon means that particles can be separated to opposite sides of the universe and a change to one is instantly reflected in the other. One possible explanation is that the concept of distance doesn't apply to such entangled particles: they are, in effect, a single entity we perceive in two separate locations.

COCKTAIL PARTY TIDBITS

❖ Even babies are aware there's something odd about action at a distance. Babies are made bored by constant repetition of a particular scene; then some small aspect of the scene is changed. If the new movement involves action with contact, the babies get less worked up than if it appears to involve action at a distance. Even to them, the whole business feels unnatural.

GRAVITY

Galileo showed that, unlike a normal force, gravity's acceleration doesn't vary with the mass being accelerated. Why is this? Because the force due to gravity involves the masses of both bodies involved (say the Earth and a brick) multiplied together. So if I work out the acceleration the brick feels, its mass from Newton's second law of motion cancels out with its mass from the gravitational equation. The brick's mass doesn't matter; it's only the mass of the Earth that's left.

Newton's gravitational equation says that you need to multiply the first mass by the second, multiply that by a constant, and divide it all by the square of the distance between the two bodies. That makes gravity an inverse square law; its strength depends on the square of the distance between the objects.

ON THE FRONTIER

The particle (boson) carrying the gravitational force is called the graviton and has never been observed. Various theories have united the other fundamental forces, but the practical theories of the present day can't pull gravity in with the other three. (Einstein spent most of the second half of his life attempting to do this and failing.) String theory, and its big brother M-theory, do offer a way to pull the forces together, but only at the cost of adding a pile of invisible dimensions to reality.

COCKTAIL PARTY TIDBITS

❖ The Ancient Greek philosopher Aristotle said that a heavier object fell faster than a light object. Such was the reverence felt for his authority that no one checked it for hundreds of years.

❖ Aristotle's idea makes sense if you compare dropping a hammer and a feather. The hammer does fall faster. But this is because the feather is slowed by air resistance. If you do this experiment on the Moon with no air resistance—as *Apollo 15* commander Dave Scott did in 1972—they hit the ground simultaneously.

ORBITS AND CENTRIFUGAL FORCE

Every time we are in a car going around a corner and get pushed in the opposite direction to the turn, we feel centrifugal force. Given that it's such a part of everyday life, it's a shame that centrifugal force doesn't exist.

Take a simple example of a theme park ride with a car on the end of an arm traveling in a circle. As the ride speeds up, you are pressed against the outer wall of the car by "centrifugal force." But what's really happening? Which of the four fundamental forces pulls you outward? Nothing is pulling you. Instead, you are experiencing Newton's first law of motion. Your natural tendency when moving is to carry on in a straight line. As you attempt to do so, the wall of the car stops you. It pulls you off the straight line into a circle. The force acting on you is inward, down the arm of the ride (this *centripetal* force does exist).

Now imagine something similar, but replace the force with gravity. Imagine you are off the surface of the planet, heading along in a straight line at a tangent to the planet's surface. Gravity pulls you toward the center of the Earth. If you are going at just the right speed, gravity constantly curves you toward the surface, but doesn't pull you in fast enough to make you crash. You are orbiting as a satellite does.

Astronauts in space float weightless. This isn't because they are so far from Earth that gravity can't be felt. If that were the case, the vessel wouldn't orbit. The reality is that everyone on board (and the ship) is in freefall toward the Earth, and so feels no gravity. Because they also move forward, they keep missing the Earth, but they are still falling.

COCKTAIL PARTY TIDBITS

❖ In a geostationary orbit, a satellite stays above the same point on the Earth; this is used for satellite TV and other communications satellites. To achieve this, satellites have to be 22,000 miles up, over the Earth's equator. Elsewhere, satellites rotating at the same speed as the Earth are called geosynchronous and regularly return to the same point, but aren't stationary. (GPS satellites use a geosynchronous orbit passing the same spot twice a day.)

ELECTROMAGNETISM

Apart from gravity, the most familiar force is electromagnetism. When you sit in a chair, for instance, the electromagnetic force between the atoms in the chair and the atoms in your body keeps you up.

Electromagnetism is the action between charged particles, either negatively charged (as with electrons) or positively charged (as with protons). Two particles with the same charge repel; two with the opposite charge attract. As atoms have an outer layer of negatively charged electrons, when a surface of atoms is pushed against another surface, the two sets of electrons repel. The atoms don't sink into each other; the top set floats a tiny distance above the bottom set. This is what happens when you sit in the chair. Your atoms float just above the atoms of the chair.

Some atoms are prone to losing external electrons or gaining extra ones, ending up with versions of the atom called an ion that is positively or negatively charged. When this occurs, the opposite charges can attract, leading to the formation of chemical compounds. So, for example, positively charged sodium ions mate up with negatively charged chlorine ions to make sodium chloride, salt. Any solid object has bonds of electromagnetic attraction holding it together, to keep it solid.

The particle that carries the electromagnetic force is the photon. When, for instance, you sit on a chair, a whole stream of photons flies between the electrons in the chair and the electrons in your body, carrying the force. Photons also travel between the electrons in every atom and its nucleus. These "internal" photons don't come out to reach your eyes, but are present all the time in countless billions. Your body, for example, is filled with light.

COCKTAIL PARTY TIDBITS

❖ Michael Faraday, who first suggested the intimate link among electricity, magnetism, and light, is said to have gone public when a colleague had an attack of nerves. On Friday, April 10, 1846, Charles Wheatstone was due to lecture at the Royal Institution in London. Seconds before, Wheatstone (a nervous speaker) ran off, leaving Faraday, who organized the lectures, to pick up the pieces. He covered Wheatstone's topic too quickly, and then had to extemporize, coming up with his remarkable original thinking. At least, that's the legend, although records suggest Wheatstone was never scheduled to speak.

STATIC ELECTRICITY

Humans first became aware of electricity as static electricity. This is an accumulation of electrical charge on the surface of an object, which can cause things to move with no obvious force, or send a spark across a gap. Static electricity can be a buildup of electrons in one place, causing a negative charge, or a shortage of electrons, producing a positive charge.

When you rub a balloon against your hair the triboelectric effect takes place (really just "electricity by rubbing"). It's reasonably easy for electrons to leave your hair, and the balloon's rubber is relatively good at adding electrons. The balloon gets a negative charge and your hair a positive charge. Your hair is attracted to the balloon, so it rises, but also each individual hair is positively charged, so they repel each other: your hair frizzes out.

We still aren't absolutely sure why the biggest static electric phenomenon, lightning, occurs. It's known that positive and negative charges separate in thunderclouds, but not why. Once there is a clear charge in the cloud it produces an opposite charge in the location where the lightning will eventually strike. This happens by induction: a strong negative charge, for instance, pushes away the electrons in the adjacent material, leaving it positively charged. When the charge difference is high enough, channels in the air ionize, forming a positive channel in the air through which the bolt of lightning flows. The average lightning bolt carries around 40,000 amps of current (a 100-watt lightbulb uses less than 1 amp). The result of the lightning streaking through the air is to slam the temperature up to around 10,000°C (18,000°F). The air escaping this region, suddenly hotter than the surface of the Sun, causes thunder.

COCKTAIL PARTY TIDBITS

- ❖ The sixth-century BC Greek philosopher Thales wrote about static electricity. He noticed that amber attracts light objects when it is rubbed. The words electric and electricity come from *electrum*, the Greek word for amber.
- ❖ Air isn't a good conductor; to break down its resistance to electricity it needs a voltage of around 75,000 volts per inch.

ELECTRICAL CURRENTS

Static electricity is fun, but it can't be used to power a TV or light your house. The electricity in everyday technology is current electricity, electricity that flows.

Current requires two things: a completed circuit and an electric potential to make things happen. We talk about an electrical current, like a current flowing in water, but electricity is not like a conventional fluid. That's why we don't have to worry about stopping up electrical sockets to prevent the electricity dripping out. Without a completed circuit connecting positive and negative, no electricity flows.

Another difference is that an electrical current is not just electrons flowing through a wire like water pumped through a pipe. The movement of electrons in a current is surprisingly lazy. The moving electrons generate photons traveling at light speed (these carry the signal when we send a message down a wire), but the electrons drift along slowly. In a copper wire with domestic levels of current, this can be as slow as a millimeter a second, but as there are many billions of them, they add up to a significant current.

Driving the current is a potential difference, a difference between the electrical potential to create a force of two points, measured in volts. This is like the difference between the top of a hill and the bottom of a hill, where the potential energy difference is what gives an object the impetus to roll down.

The reason metals are good conductors is that the structure of the metal leaves a scattering of electrons free to flow around. In a piece of wire, these electrons are always jiggling from place to place with their thermal energy, but on average the same number goes in both directions, so there is no current. It's only under the influence of a potential difference that the electrons are herded into traveling in a particular direction.

COCKTAIL PARTY TIDBITS

❖ Electrical current goes backward. When electrical diagrams were first drawn, they arbitrarily showed it flowing from positive to negative. We now know the current flows the other way, but it's too late to change the convention.

❖ Metals get less conductive as they heat up and the electrons' thermal action becomes more significant.

MAGNETS

Philosophers were aware of natural magnets (lodestones) in Ancient Greece, and treatises were written on them in the Middle Ages. The two key features of magnets for those early observers were their ability to attract, and the way a suspended magnet lined itself up in a particular direction.

These were "permanent magnets," pieces of metal that attracted other metals, such as iron. A permanent magnet has two distinct ends. If the same ends of two magnets are brought together they repel each other. Different ends attract. It was soon realized that the Earth was a huge magnet, and this was why magnetized compass needles aligned themselves in a particular direction, giving the ends of the magnet—the north and south poles—their names. Confusingly, because opposite poles attract, the north pole of the Earth is a magnetic south pole, and vice versa.

Atoms with odd sets of electrons have an unbalanced magnetic effect; they are, in essence tiny magnets, which form groups called domains. But in an unmagnetized piece of magnetic material these domains point in all directions, with no overall effect. When another magnet is used to magnetize the material it lines up the domains.

An electrical current also generates a magnetic field. Electromagnets can be more powerful than permanent magnets, and more useful because the strength of the magnetism can be varied. When an electrical charge moves through a magnetic field its path is altered. This is how traditional TVs work: a stream of electrons passes between electromagnets, which switch the direction of the particles to "draw" images on the phosphorescent screen at the front.

An obvious question raised by permanent magnets is why the magnetism doesn't run out. If we use the example of a magnet deflecting a stream of electrons, because of the angle between the force that's applied to the electrons and the direction they travel in, no work is done. This is different from electrical attraction, which does involve work.

ON THE FRONTIER

All observed magnets have both poles, but many modern theories suggest that magnetic monopoles should exist. Despite many searches they are yet to be detected.

COCKTAIL PARTY TIDBITS

❖ The interaction between a magnet and the Earth's magnetic field is used by some satellites, which energize magnetic coils to make small adjustments to the satellite's alignment.

STRONG NUCLEAR FORCE

As the name suggests, the strong force is, well, strong, about 100 times stronger than electromagnetism and 10 trillion times stronger than the weak force. As for that puny beast gravity, the strong force beats it by a factor of 1 with 38 zeroes after it.

The strong force keeps things together on the nuclear level. We never see quarks, the fundamental particles making up protons and neutrons in the atomic nucleus, on their own, as the strong force is so powerful we can't split them. Unlike every other force, the strong force doesn't get weaker as the things being attracted get farther away.

Just as electromagnetism is transmitted by bosons called photons, the strong force is thought to be transmitted by other bosons, called gluons. It's only a side effect of these gluons, the residual strong force, that produces the most obvious aspect of the strong force.

If you think about it, there's something not right about the atomic nucleus. We know it contains a number of positively charged protons. Those protons, crammed together, should fly apart with electromagnetic repulsion. But this residual strong force, a sort of leak of the gluons holding the quarks together, is strong enough to overcome the repulsion.

Just as quantum electrodynamics describes the way that electromagnetic interactions take place by exchanging photons, a parallel process is responsible for quarks exchanging gluons, called quantum chromodynamics. This name reflects the naming of the gluons after the primary colors, red, green, and blue. There is no suggestion that these particles are colored, however.

COCKTAIL PARTY TIDBITS

❖ Richard Feynman, the U.S. physicist who devised quantum electrodynamics, was scathing about the term "color." In his book *QED* he says, "The idiot physicists, unable to come up with any wonderful Greek words anymore, call this type of polarization by the unfortunate name of 'color,' which has nothing to do with color in the normal sense."

WEAK NUCLEAR FORCE

THE BASICS

The weak force can feel like a necessary but unexciting extra. It weighs in between electromagnetism (it's about 1 trillion times weaker) and gravity, which it still manages to be beat by a factor of 1,000,000,0000,000,000,000,000,000.

It is, frankly, a touch obscure. It is involved when quantum particles decay, resulting in the production of other particles. The best-known such decay is beta decay. This is where a neutron breaks down to produce a proton, an electron, and a chargeless and almost undetectable particle called an antineutrino. The electron, shooting off, is the "beta particle" that gives the process its name.

The special trick up the sleeve of this weak interaction is that it's the only one of the fundamental forces that can convert one quark into another. A neutron has an up quark and two downs. A proton has a down quark and two ups; the weak force converts a down quark into an up.

ON THE FRONTIER

The bosons carrying the weak force are different from the rest. Electromagnetism's carrier particle, the photon, has no mass and lasts indefinitely if it doesn't interact with another particle. But the carrier particles of the weak nuclear force are the heavyweight W and Z bosons.

Of these, the W boson, discovered first, is around 100 times as heavy as a proton (similar to an atom of iron) and has the same charge as an electron (there are positive and negative versions). The Z boson is a little heavier, but has no charge, which could be why it's called a Z (zero charge) boson, although it has been suggested it's the Z-list-celebrity boson.

As well as their remarkable weight, W and Z bosons don't last long: they hang around for about 3×10^{-25}th (that's 3/1 followed by 25 zeros) of a second.

COCKTAIL PARTY TIDBITS

❖ Beta decay would delight the ancient alchemists. Because it changes the number of protons in the nucleus, it turns one element into another. For instance, the radioactive substance cesium 137 undergoes beta decay to become barium.

RADIOACTIVITY

Radioactivity was discovered by Frenchman Antoine Henri Becquerel in 1896. He left some salts of uranium on a covered photographic plate and found that the plate below the salts became blackened. The uranium was spontaneously giving off energy, a phenomenon later named radioactivity. Ernest Rutherford found that the radioactive output could be split into two varieties, which he labeled alpha and beta.

Rutherford showed that the beta ray was a stream of electrons. The electrons could be made to shift off course given a suitably strong magnet. Initially the alpha ray showed no sign of bending, but eventually, with a stronger magnet, it too shifted, in the opposite direction. Both rays were renamed particles. The positively charged alpha particle was later found to be the nucleus of a helium atom. With the particles deflected out of the way, Rutherford found a third stream remained, gamma rays. These proved to be a high-energy form of light.

A radioactive source gradually decays, using up its material, a process measured by "half-life," how long it takes a lump of the material to halve its radioactive content. This varies enormously from the manmade element polonium 215, with a half-life of 0.0018 seconds to uranium 235 with a half-life of 710 million years.

Radioactivity is caused by breakdown of the atomic nucleus. The first of the three types of decay involves a heavy particle being pushed out of the nucleus. Beta decay usually involves a particle in the nucleus changing, giving off an electron or positron. And gamma decay happens when a nucleus has already partly decayed and is in a high-energy state, giving off the energy as a photon. In the radioactive process, some mass is usually converted to energy according to $E=mc^2$, blasting the particles out.

COCKTAIL PARTY TIDBITS

❖ Radioactivity was very popular when first discovered. Apart from luminous dials using radioactive paint, there were radioactive tonic drinks, radioactive toothpaste, and plenty of quack medicines based on radioactivity, assuming this was an energy-giving process.

NUCLEAR FISSION

When a radioactive nucleus splits, giving off energy, the process is termed nuclear fission. Often heavy particles such as neutrons are expelled from the nucleus, for example, in the breakdown of uranium 235, which usually produces two or three neutrons. Uranium 235 has a long half-life, so it decays slowly. But these neutrons can collide with other nuclei, triggering further fission, and so on. This is called a chain reaction.

This process is the energy source of both nuclear reactors and atomic bombs. In a power station, the chain reaction is kept at a steady stable rate. A nucleus undergoes fission, producing two or three neutrons, of which (on average) one triggers a new fission reaction, and so on. In a bomb it's forced to happen much more quickly. This results in doubling, where a pair of neutrons from the fission of one nucleus each trigger another nucleus, producing four neutrons, and so on. In the bomb, the energy is released catastrophically in a very short period of time. In the reactor the energy produced is used to heat water to make steam, which drives a turbine to generate electricity.

For nuclear weapons it is necessary to separate out uranium 235 from the more plentiful and stable uranium 238. In a reactor, however, uranium 238 is less of a problem, provided the neutrons are slowed down. Fast neutrons tend to be absorbed by uranium 238 before they can hit a 235 nucleus and trigger a chain reaction. But if they are slowed down (a process called moderating, using materials such as carbon or water) they last long enough for a steady flow of them to hit uranium 235 nuclei.

ON THE FRONTIER

There is an inherently safe design of nuclear reactor called a pebble bed reactor. In these, if the reactor gets too hot the chain reaction stops well below a dangerous temperature. It is self-regulating. There are still the difficulties of disposing of waste, but the reactor itself is in a different class of safety compared to traditional reactors.

COCKTAIL PARTY TIDBITS

❖ Fission produces two million times the energy per pound weight than gasoline does.

NUCLEAR FUSION

Although fission is the basis for current nuclear power stations, the Sun uses another process, nuclear fusion. Here two or more atomic nuclei join together and in the process give off energy.

Like fission, fusion can be used destructively. The hydrogen bomb depends on fusion for its massive impact. It contains a conventional atomic bomb, initiating a fusion reaction in another material. To enable fusion to take place, the raw materials have to be placed under extremes of temperature and pressure. This is why you don't get planet-sized stars: it takes an immense gravitational field to make it possible. The Sun uses a complex fusion reaction, taking four hydrogen nuclei to make a single helium nucleus. When undertaken artificially, "heavy water," which is a source of deuterium (hydrogen with an added neutron), is the starting point.

Fusion reactors have huge potential. The fuel is easier to obtain than uranium, and there are none of the problems of nuclear waste. But it's hard to make them work. Two technologies are being tried. In a tokomak reactor, fuel is heated to a plasma. It is kept in a powerful magnetic field to stop it touching any matter, which would be destroyed. Tokomaks are shaped like a ring donut of metal; the matter being fused is circulated around the ring by magnets, getting hotter and hotter until it fuses. The alternative is to use huge lasers to start fusion in tiny pellets of material, but this approach is still highly experimental.

ON THE FRONTIER

In 1989, researchers Martin Fleischmann and Stanley Pons claimed to have achieved cold fusion, fusion without high temperature and pressure. They sent an electrical current through heavy water, and thought that deuterium was getting so crammed together on one electrode that fusion took place. The experiment could not be repeated and the researchers were ridiculed. But since then there has been further evidence of cold fusion and work continues on it in research centers such as the U.S. Navy's Space and Naval Warfare Systems Center in San Diego.

COCKTAIL PARTY TIDBITS

❖ Ever since 1950 the prediction has been that fusion power would be 50 years away. Despite being our best hope for clean, safe power, relatively little is spent on it. At the moment there is only one working tokomak in the world, which is not big enough to be self-sustaining. The first tokomak to produce more power than it uses is expected to be operational in 2016.

CHAPTER

SIX

ENERGY

WORK AND ENERGY

We've come full circle from stuff. As $E=mc^2$ shows, matter and energy are interchangeable; energy is the fundamental component of the universe. Unlike force, which has direction as well as size (force is a vector in technical terms), energy just has size (it's a scalar).

Energy does work. Work is just energy being transferred from one place to another. When we're shoving matter around, work equates to the force you use, times the distance it moves something, although there are many other ways of doing work.

We should be wary of "laws" in physics. They were originally considered immutable principles, but some, such as Newton's laws, are approximations that only apply in limited circumstances. The closest we have to a genuine law is conservation of energy. This says that the total amount of energy in a closed system stays the same. You can't make it or lose it. But what you can do is change energy into different forms. These include potential energy, the energy stored up by lifting something up against gravity or pulling against a spring; kinetic energy, the energy of motion; heat energy,

energy from the motion of molecules in matter; and various forms of energy linked to the four fundamental forces, such as chemical (electromagnetic) and nuclear energy.

COCKTAIL PARTY TIDBITS

* Work, like energy in general, is measured in joules. It's the result of applying a force of one newton over one meter of distance. In everyday life, we often use the old-fashioned calorie. (The food calorie is really 1,000 calories.) A joule is just under a quarter of a calorie.

* The law of conservation of energy means you can't get work from nowhere, making perpetual motion machines impossible. The U.S. Patent Office got so fed up with designs for perpetual motion machines that they now no longer accept them without working models. Since then, the flow has dried up.

POWER

In ordinary English we use words such as "energy" and "power" loosely. Energy can mean the ability to get things done, and power the capability to do something. When we say person X is "powerful" we mean that she is able to make things happen. In physics, the relationship between power and energy is different.

Power, in the scientific sense, is the amount of energy transferred per second. It's the rate at which work is done. It's measured in joules per second, which are called watts. We often see kilowatts for a thousand watts, megawatts for a million watts, and gigawatts for a billion watts.

This unit is familiar around the home as we rate electrical appliances by the power they consume. You might have a 100-watt light bulb—one that burns 100 joules of energy per second—or a 900-watt microwave. For historical reasons, electricity bills aren't priced per joule, but per kilowatt-hour. Each kilowatt-hour is 3.6 million joules (3.6 megajoules).

As mechanical work is the force applied times the distance moved, and power is work divided by time, then power is also force times (distance over time)—that is, force times velocity.

The time element is crucial to power. Gasoline, for instance, delivers 15 times as much energy as the same weight of the explosive TNT. But TNT is better for blowing things up because it produces that energy in a short time. That means more power, because power is energy divided by time: the shorter the time the energy is delivered in, the more power is applied. Nuclear energy is so potent because it has millions of times as much energy per unit weight as gasoline, and (in a bomb, at least) it's faster, too.

COCKTAIL PARTY TIDBITS

❖ Horsepower is also a unit of power, usually restricted to engines. It's approximately the power a horse delivers, with a typical car having anything from 50 to 300 horsepower. One horsepower is approximately 0.75 kilowatt. This shows just how much power gasoline produces: 40 to 225 kilowatts is a lot more than the consumption of a house. Car horsepower is usually brake horsepower; this is a nominal amount the engine could produce without any load. The actual power reaching the wheels will be less.

KINETIC ENERGY

We do work to get something moving, and that energy has to go somewhere. Anything standing in the way of a heavy moving object becomes aware of the energy as a result of its motion, its kinetic energy.

The kinetic energy of a moving object is $\frac{1}{2}mv^2$, where m is the mass of the object and v is its velocity, how fast it's going. The contributory factors to kinetic energy are mass and speed, and the energy goes up much faster as you increase speed than it does as you increase mass.

You might think that the work required to get something moving would be the kinetic energy alone, but other forces are involved. The most common is friction. Not only do we have to put in enough work to produce the kinetic energy, we have to overcome the interaction between the body and whatever it is sitting on. Another real-world contributor to the energy equation is wind resistance. You can easily discover the significance of this if you try to move an open umbrella quickly.

ON THE FRONTIER

Like Newton's laws, the formula for kinetic energy is an approximation for speeds significantly slower than light speed.

As an object gets faster it's necessary to take into account relativistic factors. As the object's speed nears the speed of light, its kinetic energy heads off toward infinity.

COCKTAIL PARTY TIDBITS

❖ Wind resistance and air turbulence reduce a car engine's ability to generate kinetic energy. Although air conditioning does take energy from the engine, driving with the air conditioning off and the windows open is less efficient. Much energy is used up because of resistance from the turbulent air from the open windows.

POTENTIAL ENERGY

Potential energy is energy stored away for future use. The energy in a battery is potential energy, but the term often refers to mechanical potential energy. This is stored by raising matter to a higher location. When we lift something under the force of gravity we do work. That work adds potential energy to the lifted object, which it can then emit as it goes into motion. Bear in mind, though, that gravity isn't the only source of potential energy. When you wind a clock, you are adding potential energy to the spring.

ON THE FRONTIER

Potential energy doesn't care how the object got to the place to which it's getting. You could get a car to the top of a mountain by driving it up a long winding road, by hauling it up the side of the mountain, or by carrying it up under a helicopter. Each approach would involve different amounts of work, but the potential energy of the car sitting on the mountaintop would be the same.

Similarly, if you drove your car down the winding road and back up again, at the end of the process its potential energy would be unchanged. It purely depends on the height and the force of gravity. There's a very simple formula: it's the mass of the object times the height it's raised up, times the acceleration due to gravity, which is 9.8 meters per second per second (or about 32 feet per second per second).

COCKTAIL PARTY TIDBITS

❖ Because you have to work against gravity to give an object potential energy, gravity is like negative energy. Taking this viewpoint makes it more reasonable that all of the mass and energy in the universe can have come from practically nothing in the Big Bang. The gravitational "negative energy" of the matter in the universe offsets its mass to a great extent.

GOING PLACES: MOVEMENT, SPEED, AND VELOCITY

Movement is the natural state of everything. You might think you are sitting still, but there are fluids pumping around inside you, and the atoms in your body are constantly jiggling and twitching. Along with the Earth, you are spinning around the Earth's axis, orbiting the Sun, and traveling with the Milky Way galaxy as it blasts away from the rest of the universe.

Movement in physics terms is usually measured on axes, like those used to plot a chart. We can move in three physical dimensions, and the rate at which we move is our speed. It's just the distance we travel, divided by the time taken. We are used to measuring speed in miles per hour, but physicists use meters per second. From a physics viewpoint, however, speed isn't very interesting. What we need to know is velocity.

Velocity combines speed and direction. It's a vector quantity, one that has size and direction. Think about a helicopter that's going at 100 miles an hour forward and is rising at a rate of 50 miles an hour. We can't just add the two together and say it's going at 150 miles an hour. If we look at the direction it's moving in, it's like the third side of a triangle, where one side is 50 upwards, and the second is 100 sideways. The resultant velocity (around 112 miles an hour in this case) is the length of the third side of the triangle.

We always need to consider what velocity is relative to. Often it's reasonably obvious: we measure the velocity of a car with respect to the road. Obvious, until we meet another automobile head on. Then, the important thing is our velocity with respect to the other car. The one oddity here is light, which maintains the same velocity however you move.

* Einstein liked to play around with relativity of motion, asking if the station had arrived at the train; in relativistic terms, this makes sense.
* The word "vector" comes from a Latin word for carrying something from place to place, and "scalar," the term for something with size but no direction, is from the Latin for a ladder.

MOMENTUM

We know that a moving object requires effort to stop it from moving. This is true even in space, where the object has no weight due to gravity. The measure of the movement of an object that has mass—the property that makes it difficult it stop it or slow it down—is its momentum.

The amount of momentum something has is simply its mass multiplied by its velocity—about the simplest equation in all of physics—*mv*. Momentum, like energy, is conserved.

Momentum features in several of the key phenomena of quantum theory. One is the uncertainty principle. This says that the more detail you know about a particle's momentum, the less you know about its location, and vice versa. So if you know exactly where it is, it could have any momentum, and if you know its momentum precisely, it can be anywhere in the universe.

❖ In physics, quantities are referred to by a letter. Force, for instance, is F, and mass is m. But when physicists find a letter has already been used up, they lose the plot. It's fair enough that momentum can't be m: mass is more fundamental. Instead, momentum is p. I've yet to see an explanation why. My guess is they wanted to use o, but that was too much like a zero, so they chose the next letter.

ACCELERATION

In the real world, most things undergo acceleration. In physics we use "acceleration" to mean both acceleration and deceleration (deceleration is just negative acceleration). Acceleration is the rate at which velocity changes. Probably the best known accelerations are those of a car and the acceleration due to gravity.

A car's acceleration might be 0 to 60 in 5 seconds. This is interesting for the driver, but doesn't usefully describe how the velocity changes. What those figures show is an average acceleration of 12 miles per hour per second. It's confusing having mixed units of time. We normally use something such as feet per second per second, or in physics, meters per second per second. (That car is accelerating at 17.6 feet per second per second.)

Gravitational acceleration—falling without any wind resistance—is around 32 feet per second per second, or 9.8 meters per second per second. For each second something drops, its velocity increases by 32 feet per second.

ON THE FRONTIER

Acceleration is a change of velocity, and velocity is not just speed, but is both speed and direction. So something in orbit (in fact, anything that changes direction) is accelerating, because the direction of the velocity changes. An orbiting satellite is always changing direction, so it constantly accelerates.

COCKTAIL PARTY TIDBITS

❖ We notice acceleration when we accelerate differently from our surroundings. When a car accelerates, it begins to accelerate before your body does, so you feel pressed back into your seat. But when both surroundings and our body accelerate at the same rate, standing on the Earth, or a freefalling aircraft, we feel no effect.

A THROW OF A BALL

There are few more basic actions than throwing up a ball and catching it, yet there's plenty of interesting physics involved. The ball is given kinetic energy by your hand, but then it is subject to the force of gravity. It slows down. If the ball goes fast enough, though, it won't return to Earth; it has achieved escape velocity.

An ordinary human couldn't manage this. But for Superman to throw a ball out into space he would have to throw it at around 11,200 meters per second. Nearly seven miles a second.

Space scientists have two advantages over Superman. First, they don't send their rockets straight up. By tilting them in the direction of the Earth's rotation, they can pick up some initial speed from the Earth. More significantly, unlike the ball thrown by Superman, the rocket continues to be pushed after it leaves the surface. As long as the rocket thrust is bigger than the force of gravity, the rocket will get away.

ON THE FRONTIER

Imagine a ball that you throw straight up in the air and catch. Think of the ball at three points in its flight: immediately after it left your hand, as it hangs in the air at the top of its flight, and halfway back down to your hand. In each case, in which direction is the force on the ball? Don't worry about detail, just think of the answer in each case.

Once the ball leaves your hand there are only two forces acting on it: gravity, which is downward, and air resistance, which again acts downward. The force is down. At the top of its flight, it's natural to think everything is balanced out. But there's only a single force. Gravity. Just gravity, pulling down. Finally, as the ball falls, there is some upward force from the air resistance, but it's far outweighed by gravity. So once again the force is downward. (If you got this right, congratulate yourself. Most high school science teachers get it wrong).

COCKTAIL PARTY TIDBITS

❖ Superman's ball, thrown at escape velocity, is ten times faster than a speeding bullet.

FRICTION

Newton's first law tells us that a body in motion will stay in motion unless a force acts to stop it, yet in the real world, something moving stops unless we keep pushing. (This is why the Ancient Greeks guessed that things had a natural tendency to head for the center of the universe and stop.)

Things stop—and it's part of why they're difficult to get started in the first place—because of friction. It's much easier to move a heavy object on ice than on a concrete floor. That's because the ice is much smoother. The tiny irregularities in every surface resist movement as minuscule bumps and cracks hook together.

Friction is a nonconservative force; the work done depends on the path you take. This is different from something such as gravity, where it's just a matter of the work to get from one height to another; it's not important whether you go straight up or take a long gradual path. Work against friction converts kinetic energy into heat. On a longer path, more heat is generated, so more work is done.

ON THE FRONTIER

There are two types of friction, static friction (aka stiction) and kinetic friction. When an object is stationary, those tiny lumps and cracks in the object and the surface bed into each other, so there's significantly more resistance than when it's moving. Once you get the object in motion, it's easier to keep it going.

COCKTAIL PARTY TIDBITS

❖ Friction wastes energy, but it's handy in ordinary life. Imagine walking on a frictionless floor. It would be much worse than walking on ice: there'd literally be no grip. A frictionless glass would slip from your fingers. Without friction, we couldn't manipulate anything.

LEVERS

Archimedes apparently said, "Give me a lever long enough and a fulcrum on which to place it and I shall move the world." The lever is one of the simplest machines, yet it crops up all over the place, making use of work being force times distance. The distance between the force and the fulcrum (the balance point) produces torque. Torque is the twist applied by a particular force at a particular distance. Often you might not recognize that you are dealing with a lever. Both wheelbarrows and doors are levers: in the door, the fulcrum is the hinge, and in the wheelbarrow it's the axle.

The amount of torque generated is the amount of force at right angles to the lever times the distance away from the fulcrum. This means that in a simple lever like a seesaw, if someone sits twice as close to the fulcrum they generate half the torque, so a person who's twice the weight of her son can balance exactly if she sits half the distance from the seesaw's middle. In a wheelbarrow, the load is nearer the fulcrum than the handles, so the force required to lift it is reduced. Similarly, a door's weight averages out in the middle of the door, but the handle is near the edge, doubling the effect of your pull or push.

In some levers the force put in is *greater* than the force you get out. This may seem pointless, but it means that the speed of the point where the force comes out is greater than the speed of the part where it's input. These levers amplify velocity. An example of this is a baseball bat, where a slow twist of the hands produces a faster twist of the end of the bat. Think also of electric gates, where the motor applies its torque at the pivot.

COCKTAIL PARTY TIDBITS

❖ In the Renaissance they reckoned there were six "simple machines" of which the lever was one. It was joined by the inclined plane, the screw, the wedge, the pulley, and the wheel.

❖ Your body uses levers, whether it's in powering your arms, the strength of your jaws, or performing a push-up.

SPRINGS AND SWINGS

Two favorites of the physics world are springs and pendulums, both objects that undergo regular motion. Galileo first
considered pendulums while watching a lamp swinging on
a chain at Pisa cathedral. He noticed, timing the swinging
lamp with his pulse (possibly bored by the sermon), that
the time a pendulum took to swing wasn't linked to the distance the weight traveled. Whether it made a long stroke or
a short stroke, the time was the same. It didn't depend on
how heavy the weight was either, just the length of the
string.

Springs also provide a regular oscillating motion, as
long as they aren't pulled too far. Springs have a limit called
the elastic limit. Pull them farther than this and they trans-
form permanently. Instead of returning to their original length
when released, they deform. But when springs are kept within
the limit, they work according to a simple ratio discovered
by Robert Hooke, one of Newton's contemporaries. Hooke
discovered that the farther you stretch a spring, the more
force you get: double the stretch, double the force. Technically it's a negative force because it goes in the opposite direction to the stretch.

Pendulums were a breakthrough technology in making accurate clocks, but clockmakers soon found that pendulum arms changed in length with the room temperature, and this resulted in variation in the timing. This was overcome by using materials that don't vary much with temperature, or by using a complex pendulum called a gridiron, linking bars of different materials whose expansion counters each other.

COCKTAIL PARTY TIDBITS

❖ Robert Hooke was on the receiving end of a barbed comment from Isaac Newton. Newton wrote Hooke, "If I have seen further it is by standing on ye shoulders of giants." This sounds modest. However, Hooke had a deformed back that made him seem small in stature. No one could accuse Hooke of being a giant and it seems that Newton, whose work was criticized by Hooke, was getting his revenge.

TEMPERATURE

Temperature measures the kinetic energy in the molecules of a substance. The higher the temperature, the faster the molecules zoom around. Prior to digital thermometers, temperature was measured using the expansion and contraction of materials such as mercury or alcohol in a thin tube.

The traditional domestic temperature scale for measurements is Fahrenheit. Daniel Gabriel Fahrenheit, who devised it in the eighteenth century, used three points to fix his scale. Zero was the temperature in a bath of ice and salt, and the freezing point of water and human body temperature provided the other two points. These he set at 32 and 96 degrees, because this made the difference between the two 64, enabling him to draw gradations easily by dividing the gap in half six times.

Because of this odd origin, the Fahrenheit scale puts the temperatures of the freezing and boiling points of water at 32 and 212 degrees. (He had to tweak the scale to make the difference exactly 180 degrees.) As there are easier points to fix than a saltwater bath and human body temperature, the eighteenth-century Swedish scientist Anders Celsius came up with an alternative scale with 0 and 100 as the freezing and boiling points of water. This scale (originally Centigrade, but now Celsius) is the scientific standard.

Although scientists use Celsius, physicists often prefer a variant called the Kelvin scale. This has the same degrees as Celsius, but instead of starting at the freezing point of water, it starts at absolute zero, the coldest possible temperature. Although it's the most logical scale, it isn't practical for everyday use: a typical room temperature is around 293 K, which sounds plain wrong.

COCKTAIL PARTY TIDBITS

❖ The Kelvin scale is named after British physicist William Thompson, Lord Kelvin. Confusingly, where the units in Celsius and Fahrenheit are degrees, so a room might be at 70°F, the unit for the Kelvin scale is just the kelvin, so that water freezes at around 273 K.

HEAT

Heat is a form of energy. When two objects of different temperatures come into contact, heat (measured in joules, as with any other energy) moves from the hotter to the cooler. It's a transfer of the kinetic energy of the molecules.

There are three ways heat gets from one body to another: conduction, convection, and radiation. In conduction one set of faster-moving molecules bumps against the adjacent slower molecules and speeds them up. Some materials—metals, for instance—are better heat conductors than others. The free electrons in metals that enable them to conduct electricity carry heat through the material.

The second means for heat to get around is convection. This happens in fluids such as liquids or gases, and transfers heat from one body to another when they aren't in direct contact, via the intervening fluid. In convection, the fluid nearest the heat source becomes warmer and expands. Now less dense, it rises, carrying the heat. Although central heating uses "radiators," a lot of the heat is transferred around the house by convection. This is also how a conventional oven operates.

The final option for transfer of heat is radiation, energy transferred by electromagnetism. It's how the heat of the Sun reaches us, crossing the insulating vacuum of space.

Everything radiates heat, not just "hot" things. An object's surroundings radiate constantly too, so when an object is at the same temperature as its surroundings, the radiation it receives balances out the radiation lost.

❖ You may have heard that 50 percent (or even 75 percent) of your body heat is lost through the head. This is a myth, originating in a sales campaign for hats. The actual figure is more like 10 percent. It is worth wearing a hat, but it won't make as much difference as you might expect.

THE GREENHOUSE EFFECT

THE BASICS

The greenhouse effect is caused by water vapor and gases including carbon dioxide and methane in the atmosphere. Most of the incoming sunlight powers straight through, but infrared radiation given off by the Earth is partially absorbed by these molecules. Almost immediately the molecules release the energy again. Some continues off to space, but the rest returns to Earth, warming the surface.

Each year we pour around 26 billion tons of carbon dioxide (CO_2) into the atmosphere. Around a quarter of this is absorbed by the sea (this process is slowing down as the oceans become more acidic), and about a quarter by the land, but the rest adds to that greenhouse layer. Looking back over time (possible thanks to analysis of bubbles trapped in ancient ice cores) the carbon dioxide level was roughly stable for around 800 years until the start of the industrial revolution. Since then it has been rising.

In preindustrial times, the amount of carbon dioxide in the atmosphere was around 280 ppm (parts per million). By 2005 it had reached 380 ppm, higher than it has been at any time in the last 420,000 years.

We are used to the greenhouse effect as the bad boy of global warming, and we have pushed the effect beyond the levels at which we want it. But we shouldn't get the idea that the greenhouse effect is evil. With no greenhouse effect, the temperature would average out at –18°C (–0.4°F), 33°C (60°F) colder than it is at present. About the only living things would be bacteria.

COCKTAIL PARTY TIDBITS

❖   We only have to look into the sky at dusk or dawn when the planet
    Venus is in sight to see an out-of-control greenhouse effect. Venus is
    swathed in so much carbon dioxide that little energy gets out. It has
    average surface temperatures of 480°C (896°F)—hot enough for
    lead to run liquid—and maximum temperatures of 600°C (1112°F),
    making it the hottest planet in the solar system.

THERMAL EXPANSION

When we heat something up it expands. If you've loosened a metal screwtop lid on a jar by holding it under the hot water tap, you've used thermal expansion. The metal lid expands more than the jar, loosening its grip. For most solids, the expansion in any direction is proportional to the increase in temperature (for relatively small changes in temperature). When it cools, the reverse happens: it contracts.

When you heat up a substance the atoms or molecules in it get more energetic. They bounce around more. And that means they push away from each other, expanding the substance. Engineers have to allow for expansion. Railroad tracks can't be continuously welded from end to end of the line, or they would buckle on hot days. Similarly if you take a look at a modern bridge you will often see expansion joints to cope with movement between the bridge and the road surface.

ON THE FRONTIER

Although most things contract on cooling, as we cool water to form ice it expands, which is why it's a mistake to put a glass bottle or a can of soda into the freezer. Expansion bursts the container. Similarly as ice is heated, for the first few degrees it contracts. This action is unusual, but not unique.

For instance, acetic acid (the acid in vinegar) and silicon are both less dense as a solid than a liquid.

Water's odd behavior is caused by hydrogen bonds, attraction between the positively charged hydrogen in one water molecule and the negatively charged oxygen in another. To fit into the six-sided crystal structure of ice, these bonds have to stretch and twist, pulling the water molecules further apart.

COCKTAIL PARTY TIDBITS

❖ Hydrogen bonds mean that water molecules stick together. This makes water boil at a higher temperature than expected. Water boils at 100°C at sea level. If it weren't for hydrogen bonding, the boiling point would be below –70°C. Water wouldn't exist as a liquid on the Earth, and no water means no life.

CHANGING PHASE

When we heat or cool materials, they undergo phase changes: from solid to liquid, from liquid to gas, from gas to plasma. This phenomenon could just as easily have come in the "stuff" section, but heat plays an important role, which is why it appears here.

We know that the melting point of ice is 0°C (32°F) and the boiling point is 100°C (212°F) at sea level. But the transition from solid ice to liquid water, or from liquid water to gaseous water vapor is not instantaneous.

Usually when you heat a substance, the temperature goes up in proportion to the heat you put in, according to the material's capability to hold heat, its specific heat capacity. Imagine warming a block of ice from a few degrees below freezing. It will rise in temperature as you pour heat into it, until you reach 0°C. At this point it begins to melt. But melting involves breaking the bonds that hold the solid together, and that takes energy. So for a while, as you continue to pour heat into the ice, the temperature does not rise. It is only when the ice is converted to water that the temperature heads up again. The same thing happens on the transition from water to vapor. It doesn't matter how hard you boil water, it stays at 100°C. The energy required to make a change of state is called the latent heat.

Latent heat works both ways. When something changes from solid to liquid to gas to plasma it takes energy in. That's why a fan cools you down: it evaporates liquid off the surface of your skin, and that process sucks in heat. But going through a phase change down the chain gives off heat. This is why steam at 100°C causes worse burns than boiling water: as the steam condenses it gives off heat.

COCKTAIL PARTY TIDBITS

❖ It is possible to supercool a liquid such as water, taking it below the freezing point without it solidifying, provided there are no impurities on which a crystal can form. Water can be taken down to around −4°C. Once it is supercooled, introducing a speck of dust or a tiny amount of ice is enough to make all the water change phase extremely quickly.

THERMODYNAMICS

Thermodynamics governs the relationship between heat and work. There are four laws describing its behavior.

We start with the zeroth law, so called because it was tacked on later. The zeroth law says that two objects are in equilibrium as far as heat is concerned, if heat can flow from one to the other, but doesn't. There will be a constant flow of energy backward and forward between the two objects, but what the zeroth law really means is the net flow is zero. Another way of putting it is that if A is in equilibrium with B, and C is in equilibrium with B, then A and C are also in equilibrium.

The first law stems from the conservation of energy. It just says that the energy in a system changes to match the work it does on the outside (or is done on it), and the heat given out or absorbed. The second law is also about heat going from place to place. It says that, when left to its own devices, heat moves from a hotter part of a system to a cooler part. That sounds simple, but it has profound consequences, all the way up to the future of the universe. We come back to this in the section on entropy.

The third law says you can't get a body down to absolute zero (0 K, –273.15°C or –459.67°F) in a finite number of steps. You can always get a fraction closer, whatever temperature you are at, but you never quite make it.

COCKTAIL PARTY TIDBITS

❖ Astrophysicist Arthur Eddington said, "If someone points out to you that your pet theory of the universe is in disagreement with Maxwell's equations [the equations that describe how electromagnetism works]—then so much the worse for Maxwell's equations. If it is found to be contradicted by observation—well, these experimentalists do bungle things sometimes. But if your theory is found to be against the second law of thermodynamics I can give you no hope; there is nothing for it but to collapse in the deepest humiliation."

HEAT ENGINES

A heat engine suggests a great steam locomotive pounding away at the head of a railroad train. And a steam locomotive *is* a heat engine. It converts heat into work. That's what a heat engine is all about.

The efficiency of a heat engine is the amount of heat that gets converted into useful work. At best the efficiency is one (100 percent), but in practice there are no 100-percent-efficient heat engines. This was discovered by the nineteenth-century French engineer Nicolas Léonard Sadi Carnot. Carnot realized that no real-world heat engine would ever be totally efficient because of friction and other energy losses.

He devised an imaginary engine called a Carnot engine that had none of those real-world problems. Even then, though, the Carnot engine can't be totally efficient. The efficiency of such a heat engine is one minus the ratio of the temperatures (in Kelvin) of the heat sink (where any residual heat goes) and the heat source. This ratio is only zero (meaning the engine is totally efficient) if the heat sink is at absolute zero, which the third law of thermodynamics tells us isn't possible.

A nuclear power station is a heat engine. Nuclear fission produces heat which is used to boil water and produce superheated steam. That steam then powers a turbine, generating electricity. Heat is converted into work, making it a heat engine. The great towers we associate with nuclear power (think the Springfield Nuclear Plant in *The Simpsons*) have nothing to do with atomic energy. They are cooling towers that take the excess heat and push it out into the atmosphere. A truly efficient power station wouldn't need them.

COCKTAIL PARTY TIDBITS

❖ Thermal energy from the ground can be used as a green source of power. This involves no production of greenhouse gases (except in the manufacture of the equipment). However, it isn't highly efficient. If, for example, the ground is at 104°F and the air where we're making use of the heat is at 32°F, we can work out the best efficiency possible. We have to convert those temperatures into kelvin (313.15 and 273.15, respectively). Then the maximum efficiency is $1 - 273.15/313.15$, or $1 - 0.872$, which is 0.128, just under 13 percent efficient.

ENTROPY

Entropy is a measure of the disorder in a system: the more disorder, the higher the entropy. An alternative phrasing of the second law of thermodynamics is that entropy in a closed system stays the same or rises. If there's no input from the outside, things can get more messy, but they can't tidy themselves up.

Another way of looking at entropy is that systems run down. If applied to the universe (as far as we know, a closed system) it will eventually reach a state of chaos. However, some versions of cosmology allow for an external impact on the universe that would push it back to a lower state of entropy.

ON THE FRONTIER

The idea that entropy doesn't fall of its own accord has been used to counter evolution. The argument goes that the Earth, with its vast panoply of ordered creatures, is in a lower state of entropy than the chaotic origins in which evolutionists would have us believe. Unfortunately, a key part of the second law is "in a closed system." The Earth is not a closed system. Were it not for the vast flow of energy from the Sun, the order we see could never have formed.

It's easy to decrease entropy locally. Making a pile of bricks decreases entropy over those bricks being scattered across the floor. But to do it I have to put energy into the system.

SOUND

Sound is a familiar form of energy. When I speak, vibrations in my larynx cause air molecules to vibrate. These transmit a series of waves until the sound reaches someone's ear and those waves produce a vibration in the eardrum. A tiny amount of energy passes from my throat to someone's ear via vibrating molecules of air.

Unlike light waves, sound is a compression wave, squashing air molecules closer together, then relaxing. Sound is much slower than light, traveling at around 767 miles per hour (343 meters per second) in air at sea level. It's because it's so much slower that the sound of thunder lags behind lightning. In counting seconds to estimate how far away the storm is, we are measuring the speed of sound, ignoring light as too fast to worry about.

Sound is a wave in a material which can't exist without that material. So there is no sound in space, as there is no material in the vacuum to carry the wave.

The pitch of a sound is the frequency of the wave: the higher the pitch, the higher the frequency. A human being with good hearing can detect sound between 20 hertz (cycles per second) and 20,000 hertz, although with age we lose the higher frequencies. Most sounds are not simple waves but a whole mix of different frequencies.

In some solids, sound can be quantized, giving rise to the idea of phonons, quanta of sound. It is possible to apply quantum mechanics to the transmission of sound vibrations through a crystal lattice. This has also proved a valuable approach to understanding the way heat and electricity are conducted in bodies where electrons aren't the primary carrier. Here the energy is carried by phonons.

COCKTAIL PARTY TIDBITS

❖ Although the movie *Alien* rightly announced, "In space no one can hear you scream," many movies do accompany such explosions with sounds. They seem wrong without them. Some moviemakers leave the explosion itself silent, but we hear sound with the wave of debris that is blasted from the wreckage.

ENERGY DENSITY

The energy density of a substance, the chemical energy in its bonds released when we burn it, varies dramatically. Say I had a ton of the explosive TNT and a ton of gasoline. I set fire to each. Which generates the most energy? The gas. It has 15 times as much energy per ton as TNT. The only reason TNT is better as an explosive is that it releases its energy more quickly.

ON THE FRONTIER

Although gasoline has twice the energy density of coal, coal is 20 times cheaper for the same amount of energy. In the Second World War, the Allies successfully blockaded German oil supplies. But the Germans had a trick up their sleeve, the Fischer–Tropsch process, which converts coal into heating oil or gasoline.

The world economy is linked fundamentally to oil as a medium for storing and distributing energy. But supplies are expected to become scarcer, driving prices beyond current levels. Coal has the potential to avert this energy crisis, although it could have a disastrous impact on the environment. As oil becomes increasingly scarce, coal reserves start to look very attractive. In the United States there are between two and four billion tons of coal; this could provide enough oil for several hundred years' consumption.

Historically this process has not been used because the plants are expensive to build, and for this process to be commercial, oil would have to sell at over $50 a barrel. Even though oil prices reached $130 a barrel in 2008, we didn't see Fischer–Tropsch plants. This is partly because the process is highly polluting and also because companies are unwilling to risk the initial investment in case the oil cartels respond by slashing prices, but the process is there.

COCKTAIL PARTY TIDBITS

❖ It's the sheer quantity of energy packed into fossil fuels that explains why the airliners were able to bring down the World Trade Center on 9/11. The energy from the collisions didn't destroy the buildings: it was the energy in 60 tons of aviation fuel on each plane, the equivalent of 900 tons of TNT, that did the most damage.

❖ Weight for weight, gasoline delivers 720 times the energy of a bullet.

SOLAR ENERGY

The biggest source of energy in our neighborhood is the Sun. It keeps us warm, drives our weather system, and provides breathable air via photosynthesis. Yet we only use a tiny fraction of the energy the Sun provides. Out of the 400 billion billion megawatts produced, a mere 89 billion megawatts are available to the Earth, yet that is 5,000 times global energy consumption.

We tend to think of solar electricity generation being linked to solar panels, but sunlight can also be concentrated onto pipes to heat water or produce steam and drive a turbine. (More recent designs heat a molten mixture of nitrate salts, which can achieve temperatures as high as 600°C.) One benefit of this approach is that the energy does not have to be used immediately. The heat can be stored and used to top up the supply when sunlight is not available. A related approach uses solar troughs, curved mirrors that focus sunlight onto pipes containing fluids and again generate steam to produce electricity.

ON THE FRONTIER

The first photovoltaic cells had an efficiency of around 6 percent. Current cells work at around 20 percent efficiency, but designs will soon be commercially available that provide 40 to 50 percent efficiency, and these will be pushed above 70 percent. Arguably, though, reduction in production costs is more important. As Allen Heeger of the University of California at Santa Barbara who is working on cheap cells on plastic film has commented, "The critical comparison is dollars per watt. Even if our efficiency is lower than silicon, the cost per watt could still be better because this is such a low-cost manufacturing process."

COCKTAIL PARTY TIDBITS

❖ With direct sunlight you can get around 1 horsepower from a square yard of solar cells. This means you can't run a car on solar cells. To generate 100 horsepower would require a solar panel more than 30 yards by 30 yards on the car's roof.

❖ The output of a midsized power station can be obtained from a square kilometer of solar cells.

APPENDIX A

Physics is a big subject and there's a limit to how much you can pack into a book this size. Here is some further reading on the topics we've covered.

Atom by Piers Bizony
A good mix of biography and science as the book follows the trail that led scientists to discover just what atoms are.

Before the Big Bang by Brian Clegg
Explores the Big Bang and the alternative theories we don't hear so much of, giving a clear picture of the science behind the theories.

Before the Fallout by Diana Preston
Compelling account of the discovery of radioactivity and nuclear physics, uncovering the human stories behind the scientific discoveries.

E=mc²—A Biography of the World's Most Famous Equation
    by David Bodanis

Although not the strongest on the science itself, gives some
    excellent background to the details, the history, and the
    quirky personal aspects behind this most famous
    equation and its use.

The Fabric of the Cosmos by Brian Greene

Starts with a good introduction to relativity and quantum
    theory, then slickly takes these forward to explore the
    nature of time and the composition and origins of the
    universe.

The God Effect by Brian Clegg

The history and remarkable applications of the strangest
    phenomenon in all of physics, quantum entanglement.

Light Years by Brian Clegg

An exploration of humanity's long fascination with light,
    taking in both our ideas about what light is and the
    latest developments in light technology.

The Magic Furnace by Marcus Chown

A remarkable story taking us from the Big Bang to the
    present day in explaining from where the atoms that
    make up our universe came.

Physics for Future Presidents by Richard A. Muller
Wonderful idea of presenting the physics a U.S. president
   really should know, from atomic bombs to climate
   change.

The Quantum Zoo by Marcus Chown
Approachable introduction to quantum theory and general
   relativity.

The Trouble with Physics by Lee Smolin
Absorbing exploration of the problems with string theory
   and its failure to provide a workable alternative to the
   Standard Model.

PHYSICS ONLINE

Sources to discover more about physics online.

American Physical Society
www.aps.org

Institute of Physics
www.physics.org

Physics Central
www.physicscentral.com

Physics Classroom
www.physicsclassroom.com

Physics Today
www.physicstoday.org

Popular Science magazine
www.popsci.com

Popular Science book review site
www.popularscience.co.uk

New Scientist
www.newscientist.com

Scientific American
www.sciam.com

Fun science. Fast.

Take a "fantastic voyage" through the whorls and curves of the human brain! Learn everything from how quickly you can possibly think (and that left-handed people think faster) to why being bad feels so good (yes, there's a biochemical explanation).

Everything from quarks to galactic superclusters delivered to your eyeballs at the speed of light. Take a tour of the wonder and majesty of the cosmos, from the smallest subatomic particles to the possibility of infinite universes.